T-Labs Series in Telecommunication Services

Series editors

Sebastian Möller, Berlin, Germany
Axel Küpper, Berlin, Germany
Alexander Raake, Berlin, Germany

More information about this series at http://www.springer.com/series/10013

Michael Roland

Security Issues
in Mobile NFC Devices

 Springer

Michael Roland
School of
  Informatics/Communications/Media
University of Applied Sciences
  Upper Austria
Hagenberg
Austria

ISSN 2192-2810   ISSN 2192-2829 (electronic)
T-Labs Series in Telecommunication Services
ISBN 978-3-319-15487-9   ISBN 978-3-319-15488-6 (eBook)
DOI 10.1007/978-3-319-15488-6

Library of Congress Control Number: 2015930733

Springer Cham Heidelberg New York Dordrecht London
© Springer International Publishing Switzerland 2015
This work is subject to copyright. All rights are reserved by the Publisher, whether the whole or part of the material is concerned, specifically the rights of translation, reprinting, reuse of illustrations, recitation, broadcasting, reproduction on microfilms or in any other physical way, and transmission or information storage and retrieval, electronic adaptation, computer software, or by similar or dissimilar methodology now known or hereafter developed.
The use of general descriptive names, registered names, trademarks, service marks, etc. in this publication does not imply, even in the absence of a specific statement, that such names are exempt from the relevant protective laws and regulations and therefore free for general use.
The publisher, the authors and the editors are safe to assume that the advice and information in this book are believed to be true and accurate at the date of publication. Neither the publisher nor the authors or the editors give a warranty, express or implied, with respect to the material contained herein or for any errors or omissions that may have been made.

Printed on acid-free paper

Springer International Publishing AG Switzerland is part of Springer Science+Business Media (www.springer.com)

Preface

The recent emergence of Near Field Communication (NFC)-enabled smartphones led to an increasing interest in NFC technology and its applications by equipment manufacturers, service providers, developers, and end-users. Nevertheless, frequent media reports about security and privacy issues of electronic passports, contactless credit cards, asset tracking systems, NFC-enabled mobile phones, and proprietary contactless technologies suggest that NFC is a potentially unsafe technology whose main beneficiaries are thieves. While these weaknesses are often bound to specific applications and products, they boost the fear that NFC technology as a whole is dangerous, threatens our privacy, and helps identity theft and fraud. In order to defend their own products and services, manufacturers and service providers often position themselves on the opposite extreme, stating that their products and services incorporate sufficient countermeasures.

This book is a revised version of my Ph.D. thesis. It is written for researchers, engineers, and students interested in security aspects of mobile devices and Near Field Communication. This book contains the results of my research conducted between late 2009 and early 2013 at the NFC Research Lab Hagenberg (a research group at the University of Applied Sciences Upper Austria) in close cooperation with the Department of Computational Perception at the Johannes Kepler University Linz.

My research aims for assessing the actual state of NFC security, for discovering new attack scenarios, and for providing concepts and solutions to overcome any identified unresolved issues. Based on exemplary use-case scenarios, this work focuses on the security requirements for the interaction with NFC tags and the use of NFC card emulation. For each of these two modes of NFC, existing security concepts are identified, new attack scenarios that are possible despite these existing concepts are revealed, and solutions to overcome these issues are proposed. With the introduction of NFC to iOS (on the iPhone 6 in late 2014)—the last smartphone platform with significant market share that did not yet include NFC technology—the results of my research gained new importance.

The original thesis was finished in January 2013 and was submitted to Johannes Kepler University Linz for review in February 2013. The viva voce was successfully held in March 2013. Compared to my original thesis, this book contains updates, clarifications, and additions based on recent events.

The three years of researching, preparing, and writing this thesis were a journey with many ups and downs. I would like to thank my colleagues at the NFC Research Lab Hagenberg (Josef Langer, Christian Saminger, and Stefan Grünberger) for supporting me in many ways. I would like to thank my advisor, Josef Scharinger, and my second advisor, René Mayrhofer, for their guidance, advice, and criticism. Josef and René took the time to read this Ph.D. thesis and to provide valuable feedback. Further, I would like to thank the participants of the seminar for Ph.D. students at the Department of Computational Perception (Johannes Kepler University Linz) for giving valuable hints and starting interesting discussions. Moreover, I would also like to thank the team of First Data Austria for providing a credit card terminal for my tests.

Last, but not least, I would like to thank my family for their love and support; and I would like to thank my friends for making my life enjoyable and sociable.

Linz, Austria, December 2014   Michael Roland

Acknowledgments

This research was conducted as part of the project "4EMOBILITY" (Energy-efficient Economic and Ecological Mobility) within the EU programme "Regionale Wettbewerbsfähigkeit OÖ 2007–2013 (Regio 13)" funded by the European Regional Development Fund (ERDF) and the Province of Upper Austria (Land Oberösterreich).

Contents

1	**Introduction**		1
	1.1 Motivation		1
	1.2 Objectives		3
	1.3 Approach		4
	1.4 Contributions		5
	1.5 Publications		5
	1.6 Outline		7
	References		8
2	**Basics**		13
	2.1 Smartcards		13
		2.1.1 Protocol Stack	14
		2.1.2 Contact versus Contactless Smartcards	15
		2.1.3 Smartcard Software	17
		2.1.4 Data Structures Used on Smartcards	18
		2.1.5 PC/SC	19
	2.2 Near Field Communication		19
		2.2.1 NFC Forum	20
		2.2.2 Operating Modes	20
		2.2.3 NFC Tags	21
		2.2.4 NFC Data Exchange Format (NDEF)	22
		2.2.5 NFC Record Type Definition (RTD)	24
		2.2.6 Card Emulation	27
	2.3 EMV		28
	References		29
3	**Exemplary Use-Cases**		33
	3.1 Improving Efficiency in Automotive Environments		34
		3.1.1 Personalization in a Multi-user/Multi-car Environment	34

ix

		3.1.2	Transmission of Data Generated by Vehicle Sensors....................................	36
		3.1.3	Intelligent Cloud-Based Multimedia Applications....	38
	3.2	Generalized Use-Cases.............................		39
		3.2.1	Out-of-Band Pairing with NFC.................	39
		3.2.2	Secure Element.............................	40
	3.3	Identification of Security Aspects.....................		42
		3.3.1	Peer-to-Peer Mode..........................	42
		3.3.2	Reader/Writer Mode.........................	43
		3.3.3	Card Emulation Mode.......................	43
	References..			44
4	**Related Work**...			47
	4.1	Communication Protocol............................		47
	4.2	Flaws in Legacy Contactless Chip Card Systems..........		48
	4.3	Attacks on Contactless Smartcards.....................		49
	4.4	Security and Privacy Aspects of NFC Devices............		51
		4.4.1	Tagging and Peer-to-Peer Communication.........	52
		4.4.2	Protection for Tagging and Peer-to-Peer Communication.............................	53
		4.4.3	Integration of Secure Elements into Mobile Phones...	54
		4.4.4	Mobile Phones as Attack Platforms...............	55
	4.5	Mobile Phone and Smart Phone Security.................		56
	4.6	Combining NFC with Trusted Platform Concepts..........		59
	4.7	Flaws in Existing Mobile Wallet Implementations..........		59
	4.8	Summary.......................................		61
	References..			62
5	**Tagging**..			69
	5.1	Security Issues...................................		69
	5.2	Digital Signature for NDEF Messages...................		72
		5.2.1	Attaching a Signature to an NDEF Message........	73
		5.2.2	Maintaining Backwards Compatibility............	73
		5.2.3	Signing Individual Records....................	74
		5.2.4	Scope of a Signature.........................	74
		5.2.5	Limitations of NDEF APIs....................	77
		5.2.6	Recommended Practice.......................	78
	5.3	Establishing Trust in Digitally Signed Content............		79
		5.3.1	Public-Key Infrastructure.....................	79
		5.3.2	Mapping Content Issuer Certificates to Content.....	81
		5.3.3	Partial Signatures...........................	82
		5.3.4	Managing Content Issuer Private Keys............	84
		5.3.5	Lifespan of Certificates and Signatures...........	86

	5.4	The NFC Forum Signature RTD...................	88
		5.4.1 Signature Record	88
		5.4.2 Attaching a Signature to NDEF Messages	90
		5.4.3 Signature Coverage.......................	90
	5.5	Weaknesses of the Signature RTD	90
		5.5.1 Establishing Trust........................	91
		5.5.2 Using Remote Signatures and Certificates	91
		5.5.3 Insufficient Signature Coverage	92
		5.5.4 Record Composition Attack..................	96
	5.6	Possible Solutions to the Discovered Weaknesses..........	98
	References..................................		100
6	**Card Emulation**.............................		103
	6.1	Current Perspective on Security	103
	6.2	APIs for Access to the Secure Element	104
		6.2.1 JSR 177	105
		6.2.2 Nokia Extensions to JSR 257................	106
		6.2.3 BlackBerry...........................	107
		6.2.4 Android.............................	107
		6.2.5 Open Mobile API.......................	109
		6.2.6 Secure Element Access Control	111
		6.2.7 Comparison of Access Control Schemes	112
		6.2.8 Impact of Rooting and Jail Breaking.............	114
	6.3	New Attack Scenarios.........................	114
		6.3.1 Denial-of-Service (DoS)....................	115
		6.3.2 Software-Based Relay Attack	118
	6.4	Viability of the Software-Based Relay Attack............	122
		6.4.1 Constraints of the Protocol Layers	122
		6.4.2 Building a Card Emulator	124
		6.4.3 Prototype Implementation of the Relay System......	126
		6.4.4 Test Setup for Measurement of Communication Delays..............................	130
		6.4.5 Measurement Results.....................	135
	6.5	Possible Solutions...........................	141
	References..................................		143
7	**Software-Based Relay Attacks on Existing Applications**		147
	7.1	Google Wallet	148
		7.1.1 Preparing for an In-depth Analysis	148
		7.1.2 Static Structure........................	149
		7.1.3 Interacting with the Google Wallet On-card Component	150

		7.1.4	Google Prepaid Card: A MasterCard PayPass Card	151
	7.2		Performing a Software-Based Relay Attack	154
	7.3		Viability, Limitations and Improvements	155
		7.3.1	Getting the Relay App on Devices	156
		7.3.2	Transaction Limits	156
		7.3.3	Optimizing the Relayed Data	156
	7.4		Possible Workarounds	157
		7.4.1	Timeouts of POS Terminals	157
		7.4.2	Google Wallet PIN Verification	157
		7.4.3	Disabling Internal Mode for Payment Applets	158
	7.5		Reporting and Industry Response	159
	7.6		Analysis of the Relay-Immune Google Wallet	159
	References			160
8	**Summary and Outlook**			**163**
	8.1		Tagging	163
	8.2		Card Emulation	164
	8.3		Conclusion	166
	8.4		The Bigger Picture	166
	8.5		Future Research	167
	References			168

Appendix A: Google's Secure Element API	171
Appendix B: Modifications to Google's Secure Element API Library	175
Index	183

Acronyms

ACE	Access Control Entry
ACF	Access Control File
ACL	Access Control List
AID	Application Identifier
APDU	Application Protocol Data Unit
API	Application Programming Interface
ARA	Access Rule Applet
ARF	Access Rule File
ATC	Application Transaction Counter
ATM	Automatic Teller Machine
ATR	Answer-to-Reset
ATS	Answer-to-Select
BER-TLV	Basic Encoding Rules Tag-Length-Value Format
C-APDU	Command APDU
CA	Certification Authority
CF	Chunk Flag
CVC3	Card Verification Code 3
CVM	Cardholder Verification Method
DF	Dedicated File
DoS	Denial-of-Service
DSA	Digital Signature Algorithm
ECDSA	Elliptic Curve Digital Signature Algorithm
ECQV	Elliptic Curve Qu-Vanstone
EMV	Europay, MasterCard and Visa
FCI	File Control Information Template
FDT	Frame Delay Time
EF	Elementary File
FWT	Frame Waiting Time
GCF	Generic Connection Framework
GPS	Global Positioning System

HTTP	Hyper Text Transfer Protocol
IANA	Internet Assigned Numbers Authority
IC	Integrated Circuit
ID	Identifier
IEC	International Electrotechnical Commission
IETF	Internet Engineering Task Force
IL	ID-Length Present
IP	Internet Protocol
ISD	Issuer Security Domain
ISO	International Organization for Standardization
JAR	Java Archive
Java ME	Java Platform, Micro Edition
Java SE	Java Platform, Standard Edition
JCRMI	Java Card Remote Method Invocation
JSR	Java Specification Request
KVM	K Virtual Machine
LFSR	Linear Feedback Shift Register
LLCP	Logical Link Control Protocol
MB	Message Begin
ME	Message End
MIME	Multipurpose Internet Mail Extensions
MTM	Mobile Trusted Module
NDEF	NFC Data Exchange Format
NFC	Near Field Communication
NFC-DEP	NFC Data Exchange Protocol
ObC	On-board Credentials
OBEX	Object Exchange
OPEN	GlobalPlatform Environment
PAN	Primary Account Number
PC	Personal Computer
PC/SC	Personal Computer/Smart Card Interface
PCD	Proximity Coupling Device
PICC	Proximity Integrated Circuit Card
PIN	Personal Identification Number
PKI	Public-Key Infrastructure
POS	Point-of-Sale
PPSE	Proximity Payment System Environment
PUPI	Pseudo Unique PICC Identifier
R-APDU	Response APDU
RF	Radio Frequency
RFID	Radio Frequency Identification
RTD	Record Type Definition
SATSA	Security and Trust Services API
SDK	Software Development Kit
SE	Secure Element

SEEK	Secure Element Evaluation Kit
SHA	Secure Hash Algorithm
SIM	Subscriber Identity Module
SMS	Short Message Service
SNEP	Simple NDEF Exchange Protocol
SR	Short Record
SWP	Single Wire Protocol
SSL	Secure Sockets Layer
TCP	Transmission Control Protocol
TLS	Transport Layer Security
TLV	Tag-Length-Value Format
TNF	Type Name Format
UHF	Ultra High Frequency
UICC	Universal Integrated Circuit Card
UID	Unique Identifier
UN	Unpredictable Number
URI	Uniform Resource Identifier
URL	Uniform Resource Locator
URN	Uniform Resource Name
USB	Universal Serial Bus
VPN	Virtual Private Network
WAN	Wide Area Network
Wi-Fi	Wireless Fidelity
WTX	Frame Waiting Time Extension
XML	Extensible Markup Language

Abstract

The recent emergence of Near Field Communication (NFC)-enabled smartphones led to an increasing interest in NFC technology and its applications by equipment manufacturers, service providers, developers, and end-users. Nevertheless, frequent media reports about security and privacy issues of electronic passports, contactless credit cards, asset tracking systems, NFC-enabled mobile phones, and proprietary contactless technologies suggest that NFC is a potentially unsafe technology whose main beneficiaries are thieves. While these weaknesses are often bound to specific applications and products, they boost the fear that NFC technology as a whole is dangerous, threatens our privacy, and helps identity theft and fraud. In order to defend their own products and services, manufacturers and service providers often position themselves on the opposite extreme, stating that their products and services incorporate sufficient countermeasures.

This work aims for assessing the actual state of NFC security, for discovering new attack scenarios and for providing concepts and solutions to overcome any identified unresolved issues. Based on exemplary use-case scenarios, application-specific security aspects of NFC are extracted. The current security architectures of NFC-enabled mobile phones are evaluated with regard to the identified security aspects. As a result of the exemplary use-cases, this research focuses on the interaction with NFC tags and on card emulation. For each of these two modes of NFC, this thesis reveals attack scenarios that are possible despite existing security concepts. For the interaction with NFC tags, a new attack scenario is introduced that allows modification of tag content even though its authenticity and integrity were supposedly guaranteed by a digital signature scheme. Moreover, potential privacy issues and remaining problems have been identified in the NFC Forum's signature scheme specification. For the card emulation scenario, the mobile phone itself is identified as a significant, yet unconsidered, threat. Specifically, the well-known concept of relay attacks on smartcards is extended to the mobile phone platform. By using the phone's processing capabilities and communication facilities, relay attacks can be mounted in a significantly easier and less obvious way. These assumptions are verified through prototypical implementations. Possible solutions and workarounds to overcome these issues are outlined and evaluated with regard to their advantages and disadvantages.

Chapter 1
Introduction

1.1 Motivation

Over the last couple of years, there has been a significant change in mobile phones and their use. The trend shifted from simple mobile phones with only telephone and text messaging capabilities over feature phones to smart phones. In 2010 smart phone shipments grew 74 % over 2009 [14]. According to a survey by the market research company Nielsen [41] the smart phone market in the United States is constantly growing. In 2011, already 62 % of all mobile users aged between 25 and 34 owned a smart phone [41].

Feature phones—the first step into the direction of smart phones—were devices that featured limited additional functionality, like a camera or a music player. Current smart phones are much more than a simple telephone and pager. They are truly universal devices which perform many tasks in our everyday life. We use them as our address book, organizer, digital notepad, alarm clock, calculator, camera, photo gallery, media player, navigation system, portable web browser, e-mail client, storage device and game console. We even use them as mobile wireless access points to link other devices to the Internet. There seems to be virtually no limit to smart phone applications. A recent trend turns smart phones into clients for mobile banking and into digital wallets for mobile payment. Further, users can constantly extend the capabilities of their smart phones by installing software (so-called *apps*) through various application market places.[1] In the context of this work, the term "mobile phone" typically refers to smart phones.

A technology with the potential to further boost smart phone usability is Near Field Communication (NFC). NFC could significantly simplify many tasks around mobile phones. For instance, it can be used to quickly, easily and securely establish other wireless communication channels (like Bluetooth or Wi-Fi) between two devices. By simply tapping each other's mobile phones, users can instantly exchange content between the two devices. Users can also tap NFC-enabled objects to instantly

---

[1] Examples for application market places are Google's *Play Store* and Apple's *App store*.

retrieve Uniform Resource Locators (URLs) and other (interactive) content. Using NFC card emulation capabilities, smart phones can even act as contactless smartcards for payment, ticketing and access control applications. Thus, NFC enables devices to be secure mobile wallets that contain various credit and debit cards, tickets for public transport and events, keys for access to buildings, etc.

NFC technology was invented by NXP Semiconductors and Sony in 2002 as an evolution of their existing contactless smartcard systems [29]. Since then, the technology only became available on a limited number of mobile phones. Most of them were low-end feature phones and only a small number of them were commercially available products. It took almost a decade until NFC was introduced to a current high-end smart phone: the Google Nexus S (co-developed and manufactured by Samsung) [54]. With this step, Google not only integrated NFC technology into their "flagship" Android device, but also brought NFC functionality into their Android operating system. Thus, they laid the cornerstone for adding NFC to a whole range of Android-based smart phones by many different manufacturers. Soon, other operating system manufacturers and device manufacturers followed this trend (e.g. BlackBerry [8]). Meanwhile, less than two years after Google's initial commitment, more and more new smart phones come with NFC and the technology seems to finally become a standard feature for new devices [2, 3, 6, 7, 9, 11, 12].

However, despite all the commitment, there are still doubts about NFC throughout its stakeholders. Some device manufacturers are still unsure if the technology will have significant use-cases and, therefore, hesitate to integrate NFC into all of their new mobile phones. From an end-user's perspective there is still a lack of awareness of NFC and its applications. Users often believe that NFC is a feature only intended for "tech-savvy" people. They seem to be insufficiently educated about existing appealing applications that could potentially make their lives easier. Often, they also do not know how to interact with NFC tags and other devices (i.e. how they should tap tags and other devices with their own phone).

Also, not all aspects of NFC are completely defined. Especially NFC card emulation mode is still very blurry and leaves many unanswered questions asked by application developers and service providers. Some of the most prominent questions are:

- Who will be in control of the secure element and who will have the keys to manage it?
- Can multiple secure elements co-exist in one device?
- Who can get applications into the secure element and what are the security requirements for such applications?
- Will there be centralized entities that manage secure elements owned by different manufacturers, mobile network operators, etc.?
- Will there be standardized programming interfaces for access to the secure element?
- Is host-based card emulation a viable alternative to secure element based card emulation?

1.1 Motivation

Some of these questions have been dealt with in several scientific publications (e.g. [30–32, 35, 44]) and white papers (e.g. [15, 18, 19, 21, 23, 37, 40, 56, 57]). Nevertheless, adoption of these concepts in standards and implementations has only just begun.

Yet another issue of NFC is its security or rather its perceived security. Many people believe that NFC is a potentially unsafe technology. For instance, reports on attacks against electronic passports, contactless credit cards, asset tracking systems and NFC-compatible memory cards (e.g. NXP Semiconductors' MIFARE Classic) boost the fear that NFC technology as a whole is dangerous, threatens our privacy and helps identity theft and fraud. While news stories often exaggerate the situation, they severely damage the reputation of NFC as a whole.

In spite of all these doubts, NFC seems to finally take off and hit the mass market these days [5, 10]. Therefore, there is a huge need for verification of existing security concepts and their implementations. It has to be assured that NFC is sufficiently secure for its current and future applications. Thus, protecting end-users' privacy and safety.

For example, when users tap a touch point or a tag with their mobile phones, they expect that no harmful actions are triggered on their devices. Similarly, if users use their NFC phone as a digital wallet and store their credit cards on it, they expect that these cards are safe and secure, and that nobody else can use them without their explicit permission.

1.2 Objectives

With NFC moving into broader attention, research around this topic started to grow. In 2009, the NFC Research Lab Hagenberg (a research group of the University of Applied Sciences Upper Austria) together with VTT Technical Research Centre of Finland created the *International Workshop on Near Field Communication* as a first forum dedicated to research on all aspects of NFC. The scientific workshop covers various topics ranging from applications and services over usability and user experience to security and hardware-related research.

Security and privacy research focuses on various aspects of NFC. Among them are the interaction with smart posters and NFC tags [33, 36, 38, 39, 48, 50, 55], management of the secure element [30, 31, 33–35], security and privacy of the communication protocol itself [1, 25–28, 33], vulnerabilities of NFC-enabled mobile phones in general [58], usability of NFC-enabled mobile phones as attack platforms [16, 17, 47] and security impacts of embedding secure elements into mobile phones [51, 52].

For several years, the NFC Forum (the driving organization behind NFC) considered security only a minor priority. Their focus and, thus, their main priority was the creation of an infrastructure for applications and services. Therefore, security considerations were left to the application developers. However, recent activities by the NFC Forum suggest that this trend is starting to change. In 2010, the NFC Forum published an initial version of the *NFC Signature Record Type Definition* technical specification [4]. This specification provides a means for adding authenticity and integrity to NFC tag infrastructures (e.g. smart posters) by digitally signing the tag content. With this specification, the NFC Forum intended to make interaction with smart posters and NFC tags more secure and, therefore, reacted to security threats shown by Mulliner [38]. Meanwhile, the NFC Forum has installed the Security Technical Working Group that deals with all kinds of NFC security related topics and whose aim is to create models, specifications, guidelines and recommendations related to NFC security.

When it comes to secure element security, secure elements embedded into mobile phones and other NFC devices are usually considered to inherit all security features of the underlying smartcard architecture. Of course, a secure element also shares shortcomings and security weaknesses of regular smartcards. The mobile phone is typically regarded as an additional security feature providing capabilities that exceed those of regular plastic cards. For instance, the display and input capabilities of a mobile phone supposedly provide additional and more tamper-proof user interaction with the smartcard [22, 56].

The aim of this thesis is to assess the current state of NFC security, to discover new attack scenarios and to provide concepts and solutions to overcome any identified unresolved issues. My main research questions are:

- What are the strategies that NFC uses to provide security and privacy for its current applications, and are these measures adequate for the current applications?
- What are the main unresolved security and privacy issues of NFC?
- What steps are necessary to make NFC a reliable and secure technology?

In this thesis, I will sketch example scenarios for use-cases of NFC. These use-cases will mainly concentrate on usage of NFC in automotive environments to achieve energy-efficient economic and ecological mobility. Based on these use-cases, I will investigate the current security features and capabilities of NFC devices. The focus will lie on interaction with NFC tags and on card emulation, which seem to have the broadest use-cases at the moment. Based on the exemplary use-cases, I will extract the security requirements demanded from NFC technology and NFC devices.

1.3 Approach

This thesis assesses the existing security concepts of NFC and NFC-enabled mobile phones in the context of specific use-case scenarios. Research methods used for this assessment comprise:

- literature review,
- evaluation of protocols, standards and their existing implementations,
- characterization of attack scenarios based on exemplary use-cases,

1.3 Approach

- design of attack models and concepts based on protocol evaluation and literature review,
- design and implementation of prototypes, and
- evaluation of attack models and concepts based on these prototype implementations.

1.4 Contributions

This thesis describes exemplary use-cases of NFC and uses them as a basis for discussing the current security architectures of NFC-enabled mobile phones. As a result of the identified use-cases, this research focuses on interaction with NFC tags ("tagging" or reader/writer mode) and card emulation. For each of these two modes of NFC, this thesis reveals attack scenarios that are possible despite existing security concepts.

For the tagging scenario, I introduce an attack that allows modification of tag content even though its authenticity and integrity were supposedly guaranteed by a digital signature scheme. For the card emulation scenario, I introduce the mobile phone itself as a significant, yet unconsidered threat. Specifically, the well-known concept of relay attacks on (contactless) smartcards [25, 26, 28] is extended to the mobile phone platform. By using the processing capabilities and communication facilities of the mobile phone that contains the secure element, relay attacks can be mounted in a significantly easier and less obvious way. These assumptions have been verified through prototypical implementations.

The results of my research on digital signatures for the tagging scenario and the identified possible solutions were presented to the NFC Forum's security technical working group and have been used to create a new and robust version of the NFC Signature Record Type Definition.

My research results on relay attacks against NFC secure elements have been used to improve existing card emulation applications by adhering to strategies shown in this thesis to circumvent possible relay attacks. Specifically, my research results on relay attacks against Google Wallet have been disclosed to Google and their wallet partners prior to releasing them to the public. Google acknowledged this responsible disclosure procedure with an entry in the "Honorable Mention" section of their *Application Security Hall of Fame* [20]. This story has also been picked up by several Austrian newspapers (e.g. derStandard [43], Die Presse [13], Futurezone [59], Oberösterreichische Nachrichten [24], ORF.at [42]).

1.5 Publications

Parts of this thesis have been previously published in peer-reviewed conference proceedings, in e-print archives, in reports and in books:

- Roland, M., Langer, J., Scharinger, J.: Applying Relay Attacks to Google Wallet. In: Proceedings of the Fifth International Workshop on Near Field Communication (NFC 2013). IEEE, Zurich, Switzerland (2013). DOI 10.1109/NFC.2013.6482441
- Roland, M.: Applying recent secure element relay attack scenarios to the real world: Google Wallet Relay Attack. Computing Research Repository (CoRR), arXiv:1209.0875 [cs.CR] (2012). URL http://arxiv.org/abs/1209.0875
- Roland, M.: Software Card Emulation in NFC-enabled Mobile Phones: Great Advantage or Security Nightmare? In: 4th International Workshop on Security and Privacy in Spontaneous Interaction and Mobile Phone Use. Newcastle, UK (2012). URL http://www.medien.ifi.lmu.de/iwssi2012/papers/iwssi-spmu2012-roland.pdf
- Roland, M., Langer, J., Scharinger, J.: Relay Attacks on Secure Element-enabled Mobile Devices: Virtual Pickpocketing Revisited. In: Information Security and Privacy Research, *IFIP AICT*, vol. 376/2012, pp. 1–12. Springer, Heraklion, Creete, Greece (2012). DOI 10.1007/978-3-642-30436-1_1
- Roland, M., Langer, J., Scharinger, J.: Practical Attack Scenarios on Secure Element-enabled Mobile Devices. In: Proceedings of the Fourth International Workshop on Near Field Communication (NFC 2012), pp. 19–24. IEEE, Helsinki, Finland (2012). DOI 10.1109/NFC.2012.10
- Roland, M.: Security & Privacy Issues of the Signature RTD. Report to the NFC Forum Security Technical Working Group (2012). URL http://www.mroland.at/fileadmin/mroland/papers/201202_SignatureRTD_Security_Issues.pdf
- Roland, M., Langer, J., Scharinger, J.: Security Vulnerabilities of the NDEF Signature Record Type. In: Proceedings of the Third International Workshop on Near Field Communication (NFC 2011), pp. 65–70. IEEE, Hagenberg, Austria (2011). DOI 10.1109/NFC.2011.9
- Roland, M., Langer, J., Bogner, M., Wiesinger, F.: NFC im Automobil: Software bringt Ökonomie und braucht Sicherheit. In: L. Höfler, J. Kastner, T. Kern, G. Zauner (eds.) Energieeffiziente Mobilität, Informations- und Kommunikationstechnologie, pp. 112–119. Shaker, Aachen (2010)
- Langer, J., Roland, M.: Anwendungen der Near Field Communication Technologie und deren Nutzung in Mobiltelefonen. In: Wireless Communication and Information: Car to Car, Sensor Networks and Location Based Services, pp. 75–84. Hülsbusch, Boizenburg (2010)
- Langer, J., Roland, M.: Anwendungen und Technik von Near Field Communication (NFC). Springer Berlin Heidelberg (2010)
- Roland, M., Langer, J.: Digital Signature Records for the NFC Data Exchange Format. In: Proceedings of the Second International Workshop on Near Field Communication (NFC 2010), pp. 71–76. IEEE, Monaco (2010). DOI 10.1109/NFC.2010.10

1.6 Outline

Figure 1.1 shows the outline of this book. The remainder of this book is divided into seven chapters. Chapter 2 gives an introduction to Near Field Communication and its use in smart phones. Chapter 3 sketches use-cases that are used as the basis for the evaluation of NFC security aspects. Chapter 4 summarizes related work and positions this thesis in comparison to existing work in the areas of NFC and smart phone security. Chapters 5, 6 and 7 comprise the two main parts of this thesis. While Chap. 5 focuses on the tagging scenario, Chaps. 6 and 7 focus on card emulation mode and the use of secure elements in mobile phones. Chapter 5 first describes the security issues of tagging and evaluates digital signatures as a security measure to provide authenticity, integrity and trust for data stored on NFC tags. The chapter further discovers severe weaknesses in the existing security standard for NFC tags and provides possible solutions. Chapter 6 evaluates existing measures to protect the secure element in NFC devices. Based on this evaluation, a group of new hypothetical attack scenarios—consisting of a denial-of-service attack and a relay attack—is characterized. The viability of the relay attack in a controlled environment is verified with a prototypical implementation. Moreover, the chapter discusses possible countermeasures against these attack scenarios. Chapter 7 verifies the applicability

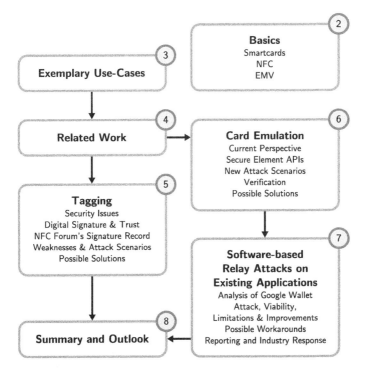

Fig. 1.1 Outline of this thesis

of the software-based relay attack scenario based on the existing payment system Google Wallet. Chapter 8 summarizes the main arguments of this thesis and provides an outlook on possible future research.

Parts of Chap. 2 have been published verbatim in [29, 45, 47, 48, 50–52]. Significant parts of Chap. 3 have been published verbatim as [49]. Parts of Chap. 4 have been published verbatim in [45, 47, 48, 50–53]. Significant parts of Chap. 5 have been published verbatim as [46, 48, 50]. Significant parts of Chaps. 6 and 7 have been published verbatim as [45, 51–53].

References

1. Anderson, R.: Position statement in RFID S&P panel: RFID and the middleman. In: Financial Cryptography and Data Security. LNCS, vol. 4886/2007, pp. 46–49. Springer, Berlin (2007). doi:10.1007/978-3-540-77366-5_6
2. Berger, P.: RIM adds two more NFC BlackBerrys. Near Field Communications World. http://www.nfcworld.com/2011/11/21/311404/ (2011)
3. Clark, M.: Virgin mobile adds gWallet phone. Near Field Communications World. http://www.nfcworld.com/2012/05/11/315619/ (2012)
4. Clark, S.: NFC Forum spec adds digital signatures to prevent tag tampering. Near Field Communications World. http://www.nfcworld.com/2010/02/11/32704/ (2010)
5. Clark, S.: 630m NFC phones in 2015. Near Field Communications World. http://www.nfcworld.com/2011/09/30/310342/ (2011)
6. Clark, S.: Acer to include NFC in all its Android phones. Near Field Communications World. http://www.nfcworld.com/2011/11/08/311164/ (2011)
7. Clark, S.: Nokia unveils N9 NFC phone. Near Field Communications World. http://www.nfcworld.com/2011/06/21/38138/ (2011)
8. Clark, S.: RIM unveils BlackBerry Bold 9900 and 9930 NFC phones. Near Field Communications World. http://www.nfcworld.com/2011/05/02/37197/ (2011)
9. Clark, S.: Samsung and Google unveil Galaxy Nexus NFC phone. Near Field Communications World. http://www.nfcworld.com/2011/10/19/310772/ (2011)
10. Clark, S.: 200m NFC phones in 2012. Near Field Communications World. http://www.nfcworld.com/2012/01/25/312711/ (2012)
11. Clark, S.: Samsung Galaxy S III expands NFC P2P capabilities with S Beam for faster file transfers. Near Field Communications World. http://www.nfcworld.com/2012/05/04/315501/ (2012)
12. Davies, J.: Hands on: The Lumia 610, Nokia's first Windows NFC phone. Near Field Communications World. http://www.nfcworld.com/2012/04/11/315025/ (2012)
13. Die Presse: Linzer Forscher löst Sicherheitsproblem für Google. DiePresse.com. http://diepresse.com/home/techscience/mobil/android/1304511/ (2012)
14. Epstein, Z.: Berg: Smartphone shipments grew 74 % in 2010. BGR. http://www.bgr.com/2011/03/10/berg-smartphone-shipments-grew-74-in-2010/ (2011)
15. European Payments Council (EPC) and GSMA: Mobile contactless payments service management roles requirements and specifications, version 2.0. Technical report EPC 220–08. http://www.gsma.com/mobilenfc/mobile-contactless-payments-service-management-roles-requirements-and-specifications-october-2010/ (2010)
16. Francis, L., Hancke, G.P., Mayes, K.E., Markantonakis, K.: Practical NFC peer-to-peer relay attack using mobile phones. In: Radio Frequency Identification: Security and Privacy Issues. LNCS, vol. 6370/2010, pp. 35–49. Springer, Berlin (2010). doi:10.1007/978-3-642-16822-2_4

17. Francis, L., Hancke, G.P., Mayes, K.E., Markantonakis, K.: Practical relay attack on contactless transactions by using NFC mobile phones. Cryptology ePrint Archive, Report 2011/618. http://eprint.iacr.org/2011/618 (2011)
18. GlobalPlatform: GlobalPlatform's Proposition for NFC Mobile: Secure Element Management and Messaging. White paper. http://www.globalplatform.org/documents/GlobalPlatform_NFC_Mobile_White_Paper.pdf (2009)
19. GlobalPlatform Mobile Task Force: Requirements for NFC mobile: management of multiple secure elements, version 1.0. Technical report GP_REQ_004. http://www.globalplatform.org/documents/whitepapers/GlobalPlatform_Requirements_Secure_Elements.pdf (2010)
20. Google: Google—Application Security—Hall of Fame—Honorable Mention. http://www.google.com/about/appsecurity/hall-of-fame/distinction/ (2014). Accessed Dec 2014
21. GSMA: Mobile NFC services, version 1.0. White paper (2007)
22. GSMA: Mobile NFC technical guidelines, version 2.0. White paper (2007)
23. GSMA: Pay-Buy-Mobile–Business opportunity analysis, version 1.0. White paper. http://www.gsma.com/mobilenfc/pay-buy-mobile-business-opportunity-analysis-november-2007/ (2007)
24. Habringer, A.: Drei Buchstaben beherrschen seine Welt. Oberösterreichische Nachrichten. http://www.nachrichten.at/oberoesterreich/art4,996318 (2012)
25. Hancke, G.P.: A practical relay attack on ISO 14443 proximity cards. http://www.rfidblog.org.uk/hancke-rfidrelay.pdf (2005). Accessed Sept 2011
26. Hancke, G.P., Mayes, K.E., Markantonakis, K.: Confidence in smart token proximity: relay attacks revisited. Comput. Secur. **28**(7), 615–627 (2009). doi:10.1016/j.cose.2009.06.001
27. Haselsteiner, E., Breitfuß, K.: Security in Near Field Communication (NFC)–strengths and weaknesses. In: Workshop on RFID Security 2006 (RFIDsec 06). Graz, Austria. http://events.iaik.tugraz.at/RFIDSec06/Program/papers/002%20-%20Security%20in%20NFC.pdf (2006)
28. Kfir, Z., Wool, A.: Picking virtual pockets using relay attacks on contactless smartcard. In: Proceedings of the First International Conference on Security and Privacy for Emerging Areas in Communications Networks (SecureComm 2005), pp. 47–58. IEEE, Athens, Greece (2005). doi:10.1109/SECURECOMM.2005.32
29. Langer, J., Roland, M.: Anwendungen und Technik von Near Field Communication (NFC). Springer, Berlin Heidelberg (2010)
30. Madlmayr, G.: A mobile trusted computing architecture for a Near Field Communication ecosystem. In: Proceedings of the 10th International Conference on Information Integration and Web-based Applications and Services (iiWAS2008), pp. 563–566. ACM, Linz, Austria (2008). doi:10.1145/1497308.1497411
31. Madlmayr, G.: Eine mobile Service Architektur für ein sicheres NFC Ökosystem. Ph.D. thesis, Johannes Kepler Universität Linz, Institut für Computational Perception (2009)
32. Madlmayr, G., Dillinger, O., Langer, J., Scharinger, J.: Management of multiple cards in NFC-Devices. In: Smart card research and advanced applications. LNCS, vol. 5189/2008, pp. 149–161. Springer, London (2008). doi:10.1007/978-3-540-85893-5_11
33. Madlmayr, G., Langer, J., Kantner, C., Scharinger, J.: NFC devices: security and privacy. In: Proceedings of the Third International Conference on Availability, Reliability and Security (ARES '08), pp. 642–647. IEEE, Barcelona, Spain (2008). doi:10.1109/ARES.2008.105
34. Madlmayr, G., Langer, J., Kantner, C., Scharinger, J., Schaumüller-Bichl, I.: Risk analysis of over-the-air transactions in an NFC ecosystem. In: Proceedings of the First International Workshop on Near Field Communication (NFC '09), pp. 87–92. IEEE, Hagenberg, Austria (2009). doi:10.1109/NFC.2009.17
35. Madlmayr, G., Langer, J., Scharinger, J.: Managing an NFC ecosystem. In: Proceedings of the 7th International Conference on Mobile Business (ICMB 2008), pp. 95–101. IEEE, Barcelona, Spain (2008). doi:10.1109/ICMB.2008.30
36. Miller, C.: Don't stand so close to me: an analysis of the NFC attack surface. Briefing at BlackHat USA. Las Vegas, NV, USA (2012)

37. Mobey Forum, Mobile Financial Services Ltd.: Mobile device security element: key findings from technical analysis version 1.0. White paper. http://www.mobeyforum.org/content/download/344/2168/file/mobey%20forum%20security%20element%20analysis%20summary%202005.pdf (2005)
38. Mulliner, C.: Vulnerability analysis and attacks on NFC-enabled mobile phones. In: Proceedings of the International Conference on Availability, Reliability and Security (ARES '09), pp. 695–700. IEEE, Fukuoka, Japan (2009). doi:10.1109/ARES.2009.46
39. Mulliner, C.: Hacking NFC and NDEF: why I go and look at it again. Talk at NinjaCon. Vienna, Austria. http://www.mulliner.org/nfc/feed/nfc_ndef_security_ninjacon_2011.pdf (2011)
40. NFC Forum: Essentials for successful NFC mobile ecosystems. White paper. http://www.nfc-forum.org/resources/white_papers/NFC_Forum_Mobile_NFC_Ecosystem_White_Paper.pdf (2008)
41. Nielsen: Generation app: 62 % of mobile users 25–34 own smartphones. Nielsenwire. http://blog.nielsen.com/nielsenwire/?p=29786 (2011)
42. ORF: Sicherheitslücke beim Bezahlen per Handy. ORF.at. http://ooe.orf.at/news/stories/2555729/ (2012)
43. Pumhösel, A.: Googles Geldtasche gehackt. derStandard.at. http://derstandard.at/1350260526386/Googles-Geldtasche-gehackt (2012)
44. Reveilhac, M., Pasquet, M.: Promising secure element alternatives for NFC technology. In: Proceedings of the First International Workshop on Near Field Communication (NFC '09), pp. 75–80. IEEE, Hagenberg, Austria (2009). doi:10.1109/NFC.2009.14
45. Roland, M.: Applying recent secure element relay attack scenarios to the real world: Google Wallet relay attack. Computing Research Repository (CoRR). arXiv:1209.0875 (cs.CR) (2012). http://arxiv.org/abs/1209.0875
46. Roland, M.: Security and privacy issues of the signature RTD. In: Report to the NFC Forum Security Technical Working Group. http://www.mroland.at/fileadmin/mroland/papers/201202_SignatureRTD_Security_Issues.pdf (2012)
47. Roland, M.: Software card emulation in NFC-enabled mobile phones: great advantage or security nightmare? In: 4th International Workshop on Security and Privacy in Spontaneous Interaction and Mobile Phone Use. Newcastle, UK. http://www.medien.ifi.lmu.de/iwssi2012/papers/iwssi-spmu2012-roland.pdf (2012)
48. Roland, M., Langer, J.: Digital signature records for the NFC data exchange format. In: Proceedings of the Second International Workshop on Near Field Communication (NFC 2010), pp. 71–76. IEEE, Monaco (2010). doi:10.1109/NFC.2010.10
49. Roland, M., Langer, J., Bogner, M., Wiesinger, F.: NFC im Automobil: Software bringt Ökonomie und braucht Sicherheit. In: Höfler, L., Kastner, J., Kern, T., Zauner, G. (eds.) Energieeffiziente Mobilität, Informations- und Kommunikationstechnologie, pp. 112–119. Shaker, Aachen (2010)
50. Roland, M., Langer, J., Scharinger, J.: Security vulnerabilities of the NDEF signature record type. In: Proceedings of the Third International Workshop on Near Field Communication (NFC 2011), pp. 65–70. IEEE, Hagenberg, Austria (2011). doi:10.1109/NFC.2011.9
51. Roland, M., Langer, J., Scharinger, J.: Practical attack scenarios on secure element-enabled mobile devices. In: Proceedings of the Fourth International Workshop on Near Field Communication (NFC 2012), pp. 19–24. IEEE, Helsinki, Finland (2012). doi:10.1109/NFC.2012.10
52. Roland, M., Langer, J., Scharinger, J.: Relay attacks on secure element-enabled mobile devices: virtual pickpocketing revisited. In: Information Security and Privacy Research, IFIP AICT, vol. 376/2012, pp. 1–12. Springer, Heraklion, Creete, Greece (2012). doi:10.1007/978-3-642-30436-1_1
53. Roland, M., Langer, J., Scharinger, J.: Applying relay attacks to Google Wallet. In: Proceedings of the Fifth International Workshop on Near Field Communication (NFC 2013). IEEE, Zurich, Switzerland (2013). doi:10.1109/NFC.2013.6482441
54. Rubin, A.: Introducing Nexus S with Gingerbread. Official Google Blog. http://googleblog.blogspot.com/2010/12/introducing-nexus-s-with-gingerbread.html (2010)

References

55. Schoo, P., Paolucci, M.: Do you talk to each poster? Security and privacy for interactions with web service by means of contact free tag readings. In: Proceedings of the First International Workshop on Near Field Communication (NFC '09), pp. 81–86. IEEE, Hagenberg, Austria (2009). doi:10.1109/NFC.2009.20
56. Smart Card Alliance Contactless Payments Council: Proximity mobile payments: leveraging NFC and the contactless financial payments infrastructure. White paper. http://www.smartcardalliance.org/resources/lib/Proximity_Mobile_Payments_200709.pdf (2007)
57. StoLPaN: Dynamic management of multi-application secure elements. White paper. http://www.nfc-forum.org/resources/white_papers/Stolpan_White_Paper_08.pdf (2008)
58. Verdult, R., Kooman, F.: Practical attacks on NFC enabled cell phones. In: Proceedings of the Third International Workshop on Near Field Communication (NFC 2011), pp. 77–82. IEEE, Hagenberg, Austria (2011). doi:10.1109/NFC.2011.16
59. Wimmer, B.: Österreicher deckt NFC-Lücke bei Google auf. Futurezone.at Technology News. http://futurezone.at/science/oesterreicher-deckt-nfc-luecke-bei-google-auf/24.586.384 (2012)

Chapter 2
Basics

This chapter summarizes basic concepts of smartcards, Near Field Communication (NFC) and payment cards.

2.1 Smartcards

Smartcards are identification cards equipped with a microchip (integrated circuit, IC). Depending on their functionality, they can be grouped into memory cards and processor cards [40]. Memory cards contain simple memory logic that can be accessed with primitive read and write commands. Processor cards contain a microprocessor that can execute complex programs. Classic smartcards have a contact interface standardized in ISO/IEC 7816-2 [22]. These contact pads can be used with various synchronous and asynchronous communication protocols. Synchronous protocols are typically used for memory cards while asynchronous protocols are typically used for processor cards. ISO/IEC 7816-3 [21] defines a set of asynchronous transmission protocols. Some smartcards even have a Universal Serial Bus (USB) interface as standardized in ISO/IEC 7816-12 [19]. Besides classical contact interfaces, cards can also have contactless interfaces that follow Radio Frequency Identification (RFID) standards.

Smartcards are present in our everyday lives. We use them as credit and debit cards, as ID cards and as access control tokens. There also exist smartcards in other form factors. For instance, the Universal Integrated Circuit Card (UICC) in mobile phones (also known as the Subscriber Identity Module (SIM) card) is a smartcard too. Even electronic passports contain contactless smartcard technology. In particular smartcards with a contactless interface are no longer bound to specific form factors. Instead they could be integrated into virtually any object.

2.1.1 Protocol Stack

The smartcard protocol stack is standardized in the ISO/IEC 7816 series. Part 1 [27] describes the physical characteristics of smartcards. Part 2 [22] describes the contact interface. Part 3 [21] describes the electrical interface and the low-level transport protocols. Part 4 [20] describes the application layer protocol.

2.1.1.1 ISO/IEC 7816-3

ISO/IEC 7816-3 [21] defines an asynchronous serial protocol for character based exchange of information between a smartcard reader ("terminal") and a smartcard. The standard further defines the reset procedure that initializes the communication with the smartcard. In response to this reset procedure, the smartcard sends its Answer-to-Reset (ATR). The ATR contains information about communication speed, a list of supported protocols and parameters and product-specific data. On top of the asynchronous serial protocol, ISO/IEC 7816-3 defines two half-duplex transport protocols: the byte-oriented protocol $T = 0$ and the block-oriented protocol $T = 1$.

2.1.1.2 ISO/IEC 7816-4

ISO/IEC 7816-4 [20] defines an application layer protocol for smartcards. The protocol consists of a file system and commands for access to the file system, management of logical communication channels, and securing the communication. The file system consists of a master file (MF), dedicated files (DFs) and elementary files (EFs). DFs can be seen as directories. They may host complete applications, group files or store data objects [20]. EFs are the leaf nodes of the file system and contain the actual data.

The application level communication protocol is mapped on top of the lower layer transport protocol (e.g. $T = 0$ or $T = 1$). Command-response pairs are called Application Protocol Data Units (APDUs). Commands are always sent from the smartcard reader to the card while responses are always sent from the card to the reader.

Table 2.1 shows a command APDU. It consists of a header and a body. The header field contains the command class (CLA), an instruction code (INS) and instruction parameters (P1, P2). The body contains data associated with the command and length fields for the command data (Lc) and the expected response (Le).

Table 2.2 shows a response APDU. It consists of a body and a trailer. The body contains the response data. The trailer contains the status word (SW1, SW2). Typical status words are:

2.1 Smartcards

Table 2.1 Command APDU (based on [20])

Field type	Field name	Size	Description
Header	CLA	1 byte	Command class
	INS	1 byte	Instruction byte
	P1	1 byte	Parameter byte 1
	P2	1 byte	Parameter byte 2
Body	Lc	0–3 bytes	Command data length N_c
	DATA	N_c bytes	Command data
	Le	0–3 bytes	Response data length N_e

Table 2.2 Response APDU (based on [20])

Field type	Field name	Size	Description
Body	DATA	$\leq N_e$ bytes	Response data
Trailer	SW1	1 byte	Status word (first byte)
	SW2	1 byte	Status word (second byte)

- `0x9000`: The command has been successfully executed.
- `0x61NN`: The command has been successfully executed, but `0xNN` bytes of further data are waiting for retrieval.
- `0x6XYY` ($2 \leq X \leq 3$): The execution of the command ended with a warning.
- `0x6XYY` ($4 \leq X$): The execution of the command ended with an error.

2.1.2 Contact versus Contactless Smartcards

Instead or in addition to a contact interface, some smartcards have a contactless interface. The most common interface for contactless smartcards is standardized in the ISO/IEC 14443 series. This standard defines a proximity RFID system based on inductive coupling with an operating frequency of 13.56 MHz. On top of the ISO/IEC 14443 protocol stack, a contactless smartcard can either use a proprietary protocol or use the APDU-based protocol defined in ISO/IEC 7816-4. Figure 2.1 shows a comparison of the protocol stack of contact-based and contactless smartcards. The ISO/IEC 14443 standard is split into four parts. Part 1 [23] specifies the physical characteristics of the card and the antenna. Part 2 [25] defines modulation and coding schemes of the bit transfer layer and the power supply of passive cards over the Radio Frequency (RF) interface. Part 3 [26] defines the activation and anti-collision sequence and a frame-based communication protocol. Part 4 [24] specifies a half-duplex block-oriented transmission protocol comparable to T = 1. ISO/IEC 14443 is split into two types: Type A and Type B. These types differ in their modulation and coding schemes, in their activation and anti-collision protocols and in their frame-

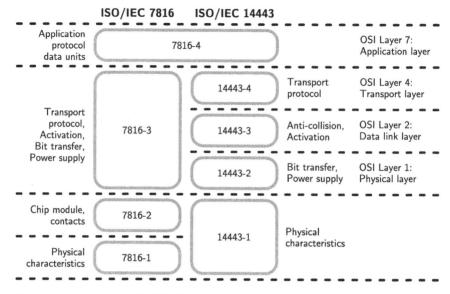

Fig. 2.1 Comparison of the ISO/IEC 7816 contact protocol stack and the ISO/IEC 14443 contactless protocol stack (*Source* [29])

based communication protocol. The block-oriented transmission protocol is the same for both types.

2.1.2.1 ISO/IEC 14443-3

ISO/IEC 14443-3 [26] is a reader-talks-first protocol. Thus, the communication is always started with a request from the reader to the card. The card then returns a response to the reader. In ISO/IEC 14443 terminology, the smartcard reader is a Proximity Coupling Device (PCD) and the smartcard is a Proximity Integrated Circuit Card (PICC).

PICCs have unique or pseudo-unique addresses that are used to identify and operate multiple cards simultaneously with one PCD. While the PCD is in idle mode, it polls for PICCs with repeated REQUEST commands (REQA for Type A and REQB for Type B).

For Type A, the REQUEST command causes all cards that have not been activated before to synchronously answer with their ATQA (Answer-to-Request). The reader then knows that at least one new card is available and continues with the anti-collision procedure. The anti-collision procedure enumerates all cards based on their Unique Identifier (UID) using a binary search tree algorithm [10]. After successful anti-collision, the PICC reveals whether it supports the transport protocol according to ISO/IEC 14443-4 or uses a proprietary transmission protocol [10].

For Type B, the REQUEST command immediately starts the anti-collision protocol which is based on a slotted-ALOHA algorithm [10]. The slotted-ALOHA algorithm works by splitting the responses of the different PICCs to the REQUEST command into multiple time slots. Each card sends its ATQB (Answer-to-Request) in its time slot. The ATQB contains the Pseudo Unique PICC Identifier (PUPI), protocol parameters and application parameters of a card. If there is a collision-free transmission in one of the time slots, the PCD has the PUPI of that PICC and can then address the card.

Besides the anti-collision and activation procedure, ISO/IEC 14443-3 also defines the frame formats and timing requirements for exchanging data between the PCD and the PICC.

2.1.2.2 ISO/IEC 14443-4

ISO/IEC 14443-4 [24] defines the protocol activation and the half-duplex block-oriented transmission protocol for contactless smartcards. For Type A, the PCD first requests the Answer-to-Select (ATS) from the PICC. The ATS is similar to the ATR of a contact-based card and contains protocol parameters and the historical bytes. The historical bytes contain free-form product identification data. For Type B, the protocol and application parameters have already been exchanged with the ATQB. After protocol activation, the transmission protocol is the same for both, Type A and B.

2.1.2.3 Other Contactless Protocols

Besides ISO/IEC 14443, also other protocols for contactless smartcards exist. ISO/IEC 15693 is a vicinity RFID standard that uses the same operating frequency (13.56 MHz) and the same communication principle (inductive coupling) as ISO/IEC 14443. However, it is specified for longer communication distances at the price of slower data rates. ISO/IEC 15693 is mainly used for simple memory cards.

Sony's proprietary FeliCa is a smartcard technology that is similar to ISO/IEC 14443. FeliCa has a file system similar to that defined in ISO/IEC 7816-4. The file system and commands for access to the file system are standardized in JIS X 6319-4 [28]. In addition, the FeliCa system has proprietary cryptography and security features.

2.1.3 Smartcard Software

Application specific smartcards (e.g. bank cards) may use customized operating systems and their application software is usually programmed into a read-only memory during the manufacturing process. Today, however, there also exist generic smartcard platforms which can be loaded with various applications. A single card can even contain multiple applications at the same time.

2.1.3.1 Java Card

A standardized framework for multi-application smartcards is the Java Card platform. Java Card operating systems provide a common set of application programming interfaces (APIs) and a standardized run-time environment. This allows development of applications that are independent of the actual smartcard hardware and of the actual operating system. As a consequence, a Java Card application that has been compiled for a certain version of the Java Card API can be run on any Java Card compliant smartcard that implements that API version.

A Java Card application consists of one or more *applets*. When an application is installed onto a Java Card each applet instance is assigned a name (application identifier, AID). Using this AID, the applet can be selected for further communication with the SELECT (file by DF name) command from the ISO/IEC 7816-4 command-set. After selection, further commands—except for the selection of other applets—are sent to the Java Card applet for processing. Hence, it is up to an applet to interpret commands and to (possibly) provide a file system like view on application data.

2.1.3.2 GlobalPlatform

Besides a common API, an application provider also needs a standardized interface to manage the lifecycle and the application software of a smartcard. "The GlobalPlatform architecture is designed to provide card issuers with the system management architecture for managing these smart cards" [13]. GlobalPlatform specifies interfaces, mechanisms and commands to allow secure smartcard application management. The management facilities are independent of the actual smartcard hardware and of the actual operating system, and are, thus, interoperable.

A GlobalPlatform compliant smartcard contains a *Card Manager*, which is the central component for card administration. It is responsible for managing card content (applications and data), security domains and the whole card lifecycle. GlobalPlatform provides standardized methods to load, install and configure applications on a smartcard. During the load operation the application executable load-file is stored on the card. Then the application and its applets can be installed, enabled for selection and personalized.

2.1.4 Data Structures Used on Smartcards

ISO/IEC 7816-4 defines five different files structures for smartcards:

1. transparent structure,
2. records of fixed size in a linear structure,

2.1 Smartcards

3. records of variable size in a linear structure,
4. records of fixed size in a cyclic structure, and
5. tag-length-value (TLV) format structure.

The transparent structure is a simple binary file format with byte-wise random access. Record files can be accessed on a per-record basis.

The TLV format is a special type of file structure where each data object consists of an identifier ("tag"), length information for the data part ("length") and a data part ("value"). TLV data objects can even be nested. Thus, the value of one data object might consist of one or more TLV data objects. TLV structures are not limited to EFs. Instead, many smartcard applications use these structures for various purposes. In this book, TLV structures are represented in the following format:

    <TAG> <LENGTH> (name)
        <VALUE> (interpretation)

For instance:

    6F 12 (FCI template)
      84 0E (DF name)
        325041592E5359532E4444463031 ("2PAY.SYS.DDF01")
      A5 00 (Proprietary information encoded in BER-TLV)

In this example 0x6F is the tag of an *FCI template* data object that contains 0x12 (18) bytes of data. The FCI template contains two nested data objects: 0x84 is the tag of a *DF name* data object with a length of 0x0E (14) bytes. The data object contains the application identifier "2PAY.SYS.DDF01" (the bytes 325041592E5359532E4444463031 represented in US-ASCII character encoding). 0xA5 is the tag of a *Proprietary information encoded in BER-TLV* data object that contains no data.

2.1.5 PC/SC

PC/SC (Personal Computer/Smart Card) is a standard to connect smartcards to PC platforms. It is supported on different operating systems (e.g. Microsoft Windows, Apple OS X and Linux). PC/SC APIs and wrapper APIs that rely on the PC/SC functionality of the underlying operating systems are available for various programming languages (e.g. C++, C#, Java and Python).

2.2 Near Field Communication

Near Field Communication (NFC) is a contactless communication technology for communication over short distances. NFC has been developed by NXP Semiconductors (formerly Philips Semiconductors) and Sony as an evolution of their inductively coupled proximity Radio Frequency Identification (RFID) technologies and smartcard technologies. NFC has originally been standardized by Ecma International in ECMA-340 [4] and ECMA-352 [3]. These standards have later been adopted by ISO/IEC in ISO/IEC 18092 [17] and ISO/IEC 21481 [18]. Further ISO/IEC and Ecma standards exist that describe test methods and enhanced interface protocols. Besides standardization through these normative bodies, further specification of protocols, data formats and NFC applications is driven by the NFC Forum.

2.2.1 NFC Forum

The NFC Forum[1] is an association of industry organizations (in particular manufacturers, application developers, and financial service institutions) and non-profit organizations with an interest in NFC. The NFC Forum was originally founded by NXP Semiconductors, Sony and Nokia to promote the use of NFC technology [29].

Today, the NFC Forum creates specifications for data formats, protocols and reference applications. A certification program based on these specifications, assures interoperability between different products and implementations.

2.2.2 Operating Modes

NFC has three operating modes:

1. peer-to-peer mode,
2. reader/writer mode, and
3. card emulation mode.

2.2.2.1 Peer-to-Peer Mode

Peer-to-peer mode is an operating mode specific to NFC and allows two NFC devices to communicate directly with each other. This mode is based on the communication protocol standardized in ISO/IEC 18092 [17]. On top of this protocol, the NFC Forum specified the Logical Link Control Protocol (LLCP) as a protocol that allows bi-directional communication between logical end-points of the two NFC devices [29]. Further high-level protocols (e.g. Simple NDEF Exchange Protocol, SNEP) allow the exchange of standardized data structures across an LLCP link.

---

[1] http://www.nfc-forum.org/.

2.2.2.2 Reader/Writer Mode

In reader/writer mode, an NFC device can access passive NFC tags. NFC tags are a subset of RFID transponders that contain a simple memory structure for storing data in a standardized format. NFC tags provide a basis for interoperability between NFC devices. Besides NFC tags, many NFC devices in reader/writer mode can also interact with other proximity RFID transponders and contactless smartcards that are based on the standards ISO/IEC 14443 [23–26] and JIS X 6319-4 [28]. Some NFC devices can even communicate with NXP's MIFARE Classic tags and with vicinity RFID transponders based on the standard ISO/IEC 15693 [14–16]. As a consequence, reader/writer mode makes NFC devices interoperable with legacy RFID tag and smartcard infrastructures.

2.2.2.3 Card Emulation Mode

In card emulation mode, an NFC device emulates a contactless smartcard. Thus, while an NFC device is in this mode, it can be accessed by existing RFID readers as if it was a regular contactless smartcard. As a consequence, card emulation mode makes NFC devices interoperable with legacy RFID reader infrastructures.

There exist several possible options for NFC card emulation mode. Emulation can differ in communication standards, in supported protocol layers, in supported command sets and in the part of the NFC device that performs the actual emulation.

With regard to the communication standard, an NFC device could emulate ISO/IEC 14443 Type A, ISO/IEC 14443 Type B or JIS X 6319-4 (Sony's FeliCa). Support for either of these modes depends on the NFC controller, the secure element and typically the geographic region. For example, ISO/IEC 14443 is the prevalent technology in Europe and North America as it is used with many payment and access control applications. For instance, contactless credit card standards are based on ISO/IEC 14443. FeliCa (JIS X 6319-4) is popular in Japan where it is used for many payment systems.

Another difference is the part of the device that performs the actual emulation. On the one hand, a contactless smartcard can be emulated by a dedicated smartcard chip, the so-called *secure element*. On the other hand, card emulation can be performed in software on the main application processor of a device (*host-based card emulation* (HCE), *software card emulation* [41], or *soft-SE* [11]).

2.2.3 NFC Tags

NFC devices in reader/writer mode and NFC tags are used to enable NFC's *tagging* application scenario. The basic principle behind tagging is "*it's all in a touch*" [2]. I.e. touching an object with an NFC device triggers an action on that device. For example, a printed advertisement could contain a tag that links to interactive content.

Table 2.3 Overview of the NFC Forum tag types (based on [29])

Tag type	RFID technology[a]	Standards	Maximum memory size
Type 1	Innovision topaz	ISO/IEC 14443-3 Type A	2 KB
Type 2	NXP MIFARE ultralight	ISO/IEC 14443-3 Type A	2 KB
Type 3	Sony FeliCa lite	JIS X 6319-4	1 MB
Type 4	NXP MIFARE DESFire[b]	ISO/IEC 14443 & ISO/IEC 7816-4	64 KB (4 GB[c])

[a]The column *RFID technology* lists only the product that the tag types were originally based on
[b]An NFC Forum Type 4 tag could be implemented on any programmable contactless smartcard that supports ISO/IEC 14443 and ISO/IEC 7816-4
[c]In version 3.0 of the Type 4 Tag Operation specification

Tapping the tag with an NFC-enabled mobile phone could cause a web site to be opened, a phone call to be initiated or a ready-made SMS message to be sent.

In order to achieve interoperability between NFC tags and NFC devices, the NFC Forum defined four different tag types that should be supported by all NFC devices. These tag types are based on existing RFID tag technologies. Table 2.3 gives an overview of the four tag types. For each tag type, the NFC Forum released a *Tag Operation Specification* [36–39]. These specifications define the memory layout of the tags and commands to access the tags.

2.2.4 NFC Data Exchange Format (NDEF)

The *NFC Data Exchange Format* (NDEF, [30]) is defined by the NFC Forum as a common format for storing data on NFC tags and for data exchange between NFC devices in peer-to-peer mode [29]. NDEF abstracts the data from the storage medium and the communication channel. Thus, on the application level an NFC device can operate on NDEF messages and need not cope with different tag platforms and operating modes.

NDEF is a simple binary data format that encapsulates application data and meta-information [29]. Data is packed into NDEF records, where each record contains type information and optional identification information for the data packet. Multiple records are grouped into one NDEF message.

2.2.4.1 NDEF Record

Figure 2.2 shows the layout of an NDEF record. A record consists of multiple header fields and a payload field. The header starts with five flag bits:

1. *Message Begin* (*MB*): MB marks the first record of an NDEF message.
2. *Message End* (*ME*): ME marks the last record of an NDEF message.
3. *Chunk Flag* (*CF*): The CF, if set, specifies that the record payload is continued in the next record.

Fig. 2.2 Layout of an NDEF record (*Source* [29])

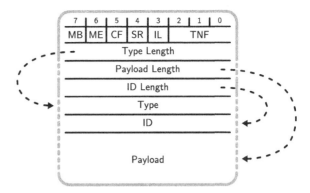

4. *Short Record* (*SR*): SR defines the size of the Payload Length field. If SR is set, the payload length is a 1-byte unsigned integer, otherwise it is a 4-byte unsigned integer. This flag is useful to reduce the memory consumption of short records.
5. *ID-Length Present* (*IL*): The IL flag, if set, specifies that the optional ID field and its corresponding length field are present.

The flags are followed by a 3-bit type classification field (Type Name Format, TNF). The value of the TNF field determines the interpretation of the Type field:

- *Empty* (0x0): The record is empty. The fields Type, ID and Payload are not present and their length fields are set to zero.
- *Well-known Type* (0x1): The Type field contains the relative Uniform Resource Identifier (URI) of an NFC Forum well-known type according to the *NFC Record Type Definition* (RTD, [31]).
- *Media Type* (0x2): The Type field contains a Multipurpose Internet Mail Extensions (MIME) media type identifier according to RFC 2046 [12].
- *Absolute URI* (0x3): The Type field contains an absolute URI according to RFC 3986 [1].
- *External Type* (0x4): The Type field contains the relative URI of an NFC Forum external type according to the *NFC Record Type Definition* (RTD, [31]).
- *Unknown* (0x5): The record contains data in an unknown format. No type information is present and the length of the Type field is zero.
- *Unchanged* (0x6): The record continues the payload of the preceding chunked record. No type information is present and the length of the Type field is zero.
- *Reserved* (0x7): This TNF value is reserved for future use.

The remaining header fields are the length information for the fields of variable length (Type Length, Payload Length, and ID Length), the type identification field (Type) and the optional record identifier (ID). The ID field may be used to specify a URI as a unique identifier for each record. This identifier can be used to cross-reference between multiple records.

Fig. 2.3 Multiple NDEF records form an NDEF message (based on [29])

The Payload field carries the actual data. The data is formatted and interpreted according to the type information in the Type field. If, for instance, the Type field specifies the MIME media type "text/x-vcard", then the payload is an electronic business card in the vCard format.

A data packet can be divided into multiple record chunks. In this case, the first record contains the type information and the optional record ID. The remaining chunks do not carry this information. Instead, their TNF field is set to 0x6 ("unchanged"). Except for the last chunk, every record chunk has its CF set, indicating that the payload is continued in the next record.

2.2.4.2 NDEF Message

Figure 2.3 shows the layout of an NDEF message. An NDEF message consists of one or more NDEF records. The first record of an NDEF message has its MB flag set. The last record has its ME flag set. The special case of an empty NDEF message is encoded by a single NDEF record with both, MB and ME set and with the TNF 0x0 ("empty").

2.2.5 NFC Record Type Definition (RTD)

The *NFC Record Type Definition* (RTD, [31]) defines two namespaces for NDEF record types: the NFC Forum well-known types and the NFC Forum external types.

NFC Forum well-known types are reserved for specifications of the NFC Forum. The type name is a Uniform Resource Name (URN) of the form "urn:nfc:wkt:<NAME>", where <NAME> identifies the type. To save storage space, the prefix "urn:nfc:wkt:" is not included into the Type field. Well-known type names can be either global or local. Global types start with an upper-case letter and have the same meaning regardless of their context. Local type names start with either a lower-case letter or a digit. They are defined for a specific context and are only valid within that context.

NFC Forum external types are reserved for self-allocation of global type names by organizations [31]. The type name is a URN of the form "urn:nfc:ext:<DOMAIN>:<NAME>", where <DOMAIN> is the issuing organizations Internet domain name and <NAME> identifies the type within that organization's namespace. In order to

2.2 Near Field Communication

Table 2.4 Format of a text record payload (based on [33])

Field	Offset[a]	Size	Coding of field content
Status byte	0	1 byte	Bit 7: Encoding of the text field (0: UTF-8, 1: UTF-16)
			Bit 6: Reserved for future use
			Bit 5..0: Length n of the language code
Language code	1	n bytes	ISO/IANA language code (US-ASCII encoded)
Text	$1+n$	m bytes	Actual text (UTF-8/UTF-16 encoded)

[a]In bytes

save storage space, the prefix "urn:nfc:ext:" is not included into the Type field. As opposed to well-known type names, external types are case-insensitive.

The NFC Forum has defined a set of well-known types. These specifications cover primitive data types as well as complex data structures for specific use-cases.

2.2.5.1 Text Record Type

The Text Record Type Definition [33] specifies a record format for free-form text with language and encoding information. The text record has the well-known type name "urn:nfc:wkt:T". Table 2.4 lists the payload format of a text record. The payload consists of a status byte, a language code and the actual text. The language code can be used to choose one out of multiple text records that best-fits a user's language preferences.

2.2.5.2 URI Record Type

The URI Record Type Definition [34] specifies a record format for URIs. Thus, the URI record can be used to store website addresses, e-mail addresses, telephone numbers, SMS messages and other information that can be represented by URIs. The URI record has the well-known type name "urn:nfc:wkt:U". Table 2.5 lists the payload format of a URI record. The payload consists of an identifier code and a URI field. The identifier code is used to reduce the size of the URI by truncating common prefixes from the URI that is stored in the URI field. Table 2.6 lists the most common identifier codes. For example, the URI "http://www.mroland.at/" could be truncated to the URI field "mroland.at/" and the identifier code `0x01`.

Table 2.5 Format of a URI record payload (based on [34])

Field	Offset[a]	Size	Coding of field content
Identifier code	0	1 byte	Prefix code for compressing the actual URI
URI field	1	n bytes	Remaining URI (UTF-8 encoded)

[a]In bytes

Table 2.6 Common identifier codes and their URI prefixes (based on [34])

Identifier code	URI prefix
0x00	No prefix, the URI field contains the full URI
0x01	"http://www."
0x02	"https://www."
0x03	"http://"
0x04	"https://"
0x05	"tel:"
0x06	"mailto:"

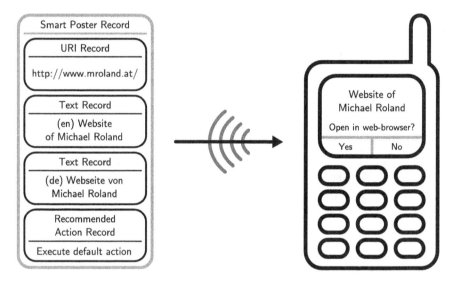

Fig. 2.4 Example of a smart poster record (based on [29])

2.2.5.3 Smart Poster Record Type

While many NFC devices associate a default action with stand-alone URI records (e.g. website addresses are opened in a web-browser), the URI Record Type Definition only specifies a container format. Instead, the NFC Forum created the *Smart Poster Record Type Definition* [32] to extend URI records with additional functionality.

A smart poster record (see Fig. 2.4) has the well-known type name "urn:nfc:wkt: Sp". Its payload is an NDEF message that consists of one URI record and optionally one or more other records. These other records can be text records that describe the URI in one or more languages, icon records that contain graphics that should be displayed together with the text, size and type information of the data referenced by the URI and a recommended action that should be performed with the URI. The Smart Poster Record Type Definition specifies three types of recommended actions:

1. Execute the default action associated with the given URI.
2. Store the URI for later use.
3. Open the URI in an appropriate editor.

Typical applications of smart poster records are advertisements with active content. Such an NFC-enabled advertisement could, for instance, link to a website or contain a ready-made SMS message for purchasing tickets [29].

2.2.5.4 Connection Handover Reference Application

The *Connection Handover* reference application [35] provides a means of using NFC as an enabler for other communication technologies (e.g. Bluetooth or Wi-Fi). The connection handover specification defines record types, message structures and handshake protocols for establishing a link through virtually any alternative carrier.

2.2.6 Card Emulation

There are several different options for performing card emulation with an NFC device. Card emulation could be performed by a secure element or by the software on the application processor ("host processor") of the device itself.

2.2.6.1 Secure Element

A secure element (SE) is a smartcard microchip that is integrated into an NFC device and connected to the NFC controller. In card emulation mode, the NFC controller routes all communication to the secure element.

A secure element can be a dedicated microchip that is embedded into the NFC device (embedded SE). Such a chip could also be combined into a single package with the NFC controller. An example for such a combined chip module is NXP's PN65N which contains a PN544 NFC controller and a secure element from NXP's SmartMX series. Another possibility is the combination of the secure element functionality with another smartcard/security device that is used within the NFC device. For instance, a UICC (also known as the SIM card) is a smartcard that is already present in many NFC devices (particularly in NFC-enabled mobile phones). Other security devices that are available equipped with smartcard technology are (micro) SD (secure digital) cards.

Many secure elements (e.g. NXP's SmartMX) are standard smartcard ICs as used for regular contact and contactless smartcards. They share the same hardware and software platforms. The only difference is the interface they provide: Instead of (or in addition to) a classic smartcard interface according to ISO/IEC 7816 (for contact cards) or an antenna (for contactless cards), the secure element has a direct interface for the connection to the NFC controller. Such interfaces are the *NFC Wired Interface* (NFC-WI, [5]) and the *Single Wire Protocol* (SWP, [9]).

2.2.6.2 Software Card Emulation

Typical use-cases for card emulation are security critical applications such as access control and payment. For these applications the secure element provides secure storage, a secure execution environment and hardware-based support for cryptographic operations. Therefore, emulation by software on a non-secure application processor was not widely used in the past. Nevertheless, several NFC controllers and NFC devices have support for software card emulation.

Software card emulation or host-based card emulation (sometimes also referred to as "soft-SE" [11]) was first made available on NFC-enabled mobile phones by BlackBerry on their BlackBerry 7 platform. A BlackBerry application can emulate NFC Forum type 4 tags and ISO/IEC 14443-4 smartcards. For an NFC Forum type 4 tag, the application simply needs to define an NDEF message that is used by the operating system to emulate the virtual tag. For an ISO/IEC 14443-4 smartcard, the application receives the command APDUs sent by the smartcard terminal and generates corresponding response APDUs.

BlackBerry mobile phones were the first mobile phones known to support software card emulation. However, patches [42, 43] to the CyanogenMod aftermarket firmware for Android devices brought software card emulation to Android devices with NXP's PN544 NFC controller. Starting with Android 4.4, software card emulation became available on most Android NFC devices under the term *host-based card emulation* (HCE). Besides that, some NFC readers (e.g. ACS ACR 122U) can be used to perform software card emulation on PC platforms.

2.3 EMV

EMV is a series of standards for chip-based credit and debit cards. The EMV standards were initially created by Europay, MasterCard and Visa (hence the acronym EMV) in an effort to design a worldwide standard for chip-based payment cards and payment terminals [8]. Today, the EMV standards are maintained by EMVCo, an organization owned by the credit card companies American Express, JCB, MasterCard and Visa [8].

The EMVCo website [8] names several advantages of EMV chip-based payment cards in comparison to chip-less cards (e.g. magnetic stripe cards):

- EMV cards prevent fraud because the smartcard chip can contain data that is close to impossible to be cloned while magnetic stripes can easily be copied.
- A smartcard is capable of computing unique digital signatures/authentication codes for each transaction in both, online and offline environments.

- Smartcards support enhanced cardholder verification methods (e.g. offline PIN verification).
- A smartcard chip has significantly more data storage than a magnetic stripe card.

The *EMV Integrated Circuit Card Specifications for Payment Systems* [7] are based on ISO/IEC 7816 smartcard standards. An EMV payment card (also referred to as *Chip & PIN* card) can support various online and offline transaction modes. Possible security measures include authenticating the card to the payment terminal, authenticating the payment terminal to the card and authenticating the cardholder to the card. Besides interaction between the terminal and the smartcard, they also define test methods, secure key management, the interface between the payment terminal and its users and the interface between the payment terminal and the bank that manages the payment.

EMVCo does not only support contact-based smartcards. Instead, they extended their scope to contactless and mobile payment systems. The *EMV Contactless Specifications for Payment Systems* [6] are a series of standards for contactless payment systems based on ISO/IEC 14443 and ISO/IEC 7816. The EMV specifications for contactless payment systems are a mere aggregation of four different payment systems (one by JCB, one by MasterCard, one by Visa and one by American Express). For each of these payment systems, the terminal has a separate software module that processes transactions.

MasterCard's EMV-compliant contactless payment system is called MasterCard PayPass. It supports two different operating modes: emulation of the magnetic stripe system (*EMV Mag-Stripe* mode) and *EMV mode*. In Mag-Stripe mode, the card stores information comparable to that on the magnetic stripe and generates dynamic authentication codes to authorize payments. As the authorization codes do not contain any information about the payment transaction, this mode is for online transactions only. In EMV mode, the card authenticates itself to the terminal and signs the payment transaction, making it possible to verify and store transactions offline for later processing.

Other EMV-compliant contactless payment systems are Visa's payWave and American Express's ExpressPay.

References

1. Berners-Lee, T., Fielding, R.T., Masinter, L.: RFC 3986: Uniform Resource Identifier (URI): Generic Syntax. Internet Engineering Task Force (2005)
2. Chen, E.: NFC: Short range, long potential. Assa Abloy FutureLab News (2007). http://www.assaabloyfuturelab.com/FutureLab/Templates/Page2Cols____1905.aspx
3. Ecma International: ECMA-352: Near Field Communication Interface and Protocol-2 (NFCIP-2) (2003)
4. Ecma International: ECMA-340: Near Field Communication Interface and Protocol (NFCIP-1) (2004)
5. Ecma International: ECMA-373: Near Field Communication Wired Interface (NFC-WI) (2006)

6. EMVCo: EMV Contactless Specifications for Payment Systems (Books A-D). Version 2.1 (2011)
7. EMVCo: EMV Integrated Circuit Card Specifications for Payment Systems (Books 1–4). Version 4.3 (2011)
8. EMVCo: About EMV (2012). http://www.emvco.com/about_emv.aspx
9. European Telecommunications Standards Institute: ETSI TS 102 613: Smart Cards; UICC—Contactless Front-end (CLF) Interface; Part 1: Physical and data link layer characteristics (Release 11) (2012)
10. Finkenzeller, K.: RFID-Handbuch, 4th edn. Carl Hanser Verlag, München (2006)
11. Francis, L., Hancke, G.P., Mayes, K.E., Markantonakis, K.: Practical Relay Attack on Contactless Transactions by Using NFC Mobile Phones. Cryptology ePrint Archive, Report 2011/618 (2011). http://eprint.iacr.org/2011/618
12. Freed, N., Borenstein, N.S.: RFC 2046: Multipurpose Internet Mail Extensions (MIME) Part Two: Media Types. Internet Engineering Task Force (1996)
13. GlobalPlatform: Card Specification. Version 2.2.1 (2011)
14. International Organization for Standardization: ISO/IEC 15693–1: Identification cards—Contactless integrated circuit(s) cards—Vicinity cards—Part 1: Physical characteristics (2000)
15. International Organization for Standardization: ISO/IEC 15693–2: Identification cards—Contactless integrated circuit(s) cards—Vicinity cards—Part 2: Air interface and initialization (2000)
16. International Organization for Standardization: ISO/IEC 15693–3: Identification cards—Contactless integrated circuit(s) cards—Vicinity cards—Part 3: Anticollision and transmission protocol (2001)
17. International Organization for Standardization: ISO/IEC 18092: Information technology—Telecommunications and information exchange between systems—Near Field Communication—Interface and Protocol (NFCIP-1) (2004)
18. International Organization for Standardization: ISO/IEC 21481: Information technology—Telecommunications and information exchange between systems—Near Field Communication Interface and Protocol-2 (NFCIP-2) (2005)
19. International Organization for Standardization: ISO/IEC 7816–12: Identification cards—Integrated circuit cards—Part 12: Cards with contacts—USB electrical interface and operating procedures (2005)
20. International Organization for Standardization: ISO/IEC 7816–4: Identification cards—Integrated circuit cards—Part 4: Organization, security and commands for interchange (2005)
21. International Organization for Standardization: ISO/IEC 7816–3: Identification cards—Integrated circuit cards—Part 3: Cards with contacts—Electrical interface and transmission protocols (2006)
22. International Organization for Standardization: ISO/IEC 7816–2: Identification cards—Integrated circuit cards—Part 2: Cards with contacts—Dimensions and location of the contacts (2007)
23. International Organization for Standardization: ISO/IEC 14443–1: Identification cards—Contactless integrated circuit cards—Proximity cards—Part 1: Physical characteristics (2008)
24. International Organization for Standardization: ISO/IEC 14443–4: Identification cards—Contactless integrated circuit cards—Proximity cards—Part 4: Transmission protocol (2008)
25. International Organization for Standardization: ISO/IEC 14443–2: Identification cards—Contactless integrated circuit cards—Proximity cards—Part 2: Radio frequency power and signal interface (2010)
26. International Organization for Standardization: ISO/IEC 14443–3: Identification cards—Contactless integrated circuit cards—Proximity cards—Part 3: Initialization and anticollision (2011)
27. International Organization for Standardization: ISO/IEC 7816–1: Identification cards—Integrated circuit cards—Part 1: Cards with contacts—Physical characteristics (2011)
28. Japanese Standards Association: JIS X 6319–4: Specification of implementation for integrated circuit(s) cards—Part 4: High Speed proximity cards (2005)

References

29. Langer, J., Roland, M.: Anwendungen und Technik von Near Field Communication (NFC). Springer, Heidelberg (2010)
30. NFC Forum: NFC Data Exchange Format (NDEF). Technical specification, version 1.0 (2006)
31. NFC Forum: NFC Record Type Definition (RTD). Technical specification, version 1.0 (2006)
32. NFC Forum: Smart Poster Record Type Definition. Technical specification, version 1.0 (2006)
33. NFC Forum: Text Record Type Definition. Technical specification, version 1.0 (2006)
34. NFC Forum: URI Record Type Definition. Technical specification, version 1.0 (2006)
35. NFC Forum: Connection Handover. Technical specification, version 1.1 (2008)
36. NFC Forum: Type 1 Tag Operation Specification. Technical specification, version 1.1 (2011)
37. NFC Forum: Type 2 Tag Operation Specification. Technical specification, version 1.1 (2011)
38. NFC Forum: Type 3 Tag Operation Specification. Technical specification, version 1.1 (2011)
39. NFC Forum: Type 4 Tag Operation Specification. Technical specification, version 2.0 (2011)
40. Rankl, W., Effing, W.: Handbuch der Chipkarten, 4th edn. Carl Hanser Verlag, München (2002)
41. Roland, M.: Software Card Emulation in NFC-enabled Mobile Phones: Great Advantage or Security Nightmare? In: 4th International Workshop on Security and Privacy in Spontaneous Interaction and Mobile Phone Use. Newcastle, UK (2012). http://www.medien.ifi.lmu.de/iwssi2012/papers/iwssi-spmu2012-roland.pdf
42. Yeager, D.: Added NFC Reader support for two new tag types: ISO PCD type A and ISO PCD type B (2012). https://github.com/CyanogenMod/android_packages_apps_Nfc/commit/d41edfd794d4d0fedd91d561114308f0d5f83878
43. Yeager, D.: Added NFC Reader support for two new tag types: ISO PCD type A and ISO PCD type B (2012). https://github.com/CyanogenMod/android_external_libnfc-nxp/commit/34f13082c2e78d1770e98b4ed61f446beeb03d88

Chapter 3
Exemplary Use-Cases

Mobile phones and Near Field Communication (NFC) have the potential to be important tools in future automotive environments. Particularly in the field of energy efficiency and economic improvements, NFC opens up for significant simplifications and can provide a good basis for securing systems.

Mobile phones become our every day universal tools. Statistics reveal that, on average, there is more than one mobile phone contract per EU citizen [4]. Especially with regard to automotive environments, mobile phones and NFC could significantly influence future applications.

At the moment, mobile phones are mainly used as minor add-ons to the automotive environment. For instance, the mobile phone is used as an in-car telephone through simple hands-free equipment. Smart phones are also often used as autonomous navigation systems. However, there is a trend towards further interconnecting the mobile phone and the automobile: In-car navigation systems can often access the mobile phone address book to browse for destination addresses. Some cars can even (automatically) call breakdown or emergency services through the hands-free equipment. In the future, the mobile phone could become the central interface between the automobile and the cellular network (and, consequently, between the automobile and the Internet).

NFC technology has several benefits that ease integration of mobile phones into automobiles. Recent research by Steffen et al. [15] shows various new possibilities for NFC in automobiles. For instance, NFC provides an easy, reliable and fast means to establish wireless connections (like Bluetooth and Wi-Fi) between the automobile and other devices (e.g. an NFC-enabled handset). This capability makes it particularly easy to link a mobile phone to an automotive computer system. Besides this "out-of-band pairing" capability, NFC with its secure element could also be used as a replacement for car keys.

While the NFC channel on its own is susceptible to eavesdropping and potentially also to message injection, its short communication distance adds a notion of explicit user interaction. I.e. by tapping an NFC tag or an NFC reader, users declare that they explicitly intend to trigger a specific action.

3.1 Improving Efficiency in Automotive Environments

A link between the computer system of an automobile and a mobile phone opens up a whole new range of possibilities for improving the efficiency of processes in and around cars. This thesis focuses on two main aspects:

- Each mobile phone is tightly bound to a specific person. Thus, it is possible to derive a context for customization of the automobile system based on the detected mobile phone.
- A mobile phone provides a permanent link to the cellular network and the Internet. In comparison to equipping a car directly with a cellular modem, the mobile phone is bound to its user's cellular network account. Thus, it is possible to easily switch to a user's preferred cellular network in a multi-user environment by using each user's mobile phone.

Three groups of applications that rely on those aspects are analyzed in detail in this thesis:

1. personalization of a car to its user in a multi-user or even a multi-car environment,
2. transmission of data generated by sensors of a vehicle, and
3. intelligent cloud-based multimedia applications.

3.1.1 Personalization in a Multi-user/Multi-car Environment

Radio Frequency Identification (RFID) technology is already widely used in automobiles. An example is car immobilizer systems: An RFID chip is embedded into each car key. The immobilizer will disable the car ignition if it does not detect a genuine RFID chip. Thus, it becomes more difficult to clone a car key.

Instead of a passive RFID transponder, an NFC-enabled mobile phone could be used to perform this task. In that case, the secure element inside the phone would contain an application that generates the codes necessary to disable the immobilizer system. A major benefit of this scenario would be that the secure element is capable of using state-of-the-art cryptography and guarantees a high level of security.

In addition to the token for the immobilizer system, an NFC-enabled mobile phone has the potential to replace the whole car key (Fig. 3.1). If a car is equipped with NFC-enabled door locks, the doors can be unlocked as soon as the mobile phone is brought into close proximity of such an NFC-enabled door lock (cf. Steffen et al. [15]).

Additionally, an NFC-based car key could allow the car to uniquely identify each user (i.e. each driver). This can be achieved with a user identifier that is unique for each user's secure element application. Therefore, keys for a car can be bound to specific users. This allows the automotive computer system to distinguish between multiple drivers upon unlocking the car and throughout its use. As a consequence, a car can adapt to its driver. For instance, parameters like the position of the seat and the mirrors, the air-condition setup and settings of the car radio could be automatically

3.1 Improving Efficiency in Automotive Environments

Fig. 3.1 An NFC-enabled mobile phone can act as car keys and additionally provide driver identification (based on [14])

adjusted to each individual driver (cf. Steffen et al. [15]). An automobile could, therefore, "learn" each driver's individual profile, associate it with the driver's NFC key and recall that profile whenever the driver's NFC key is used.

In comparison to an identification solution that uses only Bluetooth to detect the current driver, an NFC-based solution has the advantage that the driver needs to bring the NFC key in close proximity to a touch point and, therefore, expresses an explicit intention. In a multi-user environment, this assures that only the person that actually intended to claim the car is detected as the current driver. For a Bluetooth-based system, it might be problematic to detect the current driver if multiple persons inside the car would qualify as drivers (i.e. if multiple persons had Bluetooth-enabled car keys that would allow them to use that particular car).

Besides parameters for the comfort of the driver, future automobiles could also optimize parameters of the engine control unit to each driver's individual profile to increase the energy efficiency of the car (cf. BMW's project Ilena [6]). Start-stop systems or kinetic energy recovery systems could use the operational profile of a car to adapt to each driver's habits. Especially hybrid cars could gain valuable information to optimize the use of available energy [6].

Another application of per-user operational profiles is the evaluation of that data with regard to acceleration and deceleration characteristics. This information could be used to provide the drivers with feedback and to suggest improvements of their driving habits.

Organizations owning multiple vehicles that are shared by multiple users (i.e. large companies as well as car-sharing providers and car-rental companies) could particularly benefit from NFC-based car key solutions (cf. Steffen et al. [15]). Permissions for access to individual vehicles can be managed online through a centralized system. Thus, an NFC-based system could significantly simplify and optimize shared use of cars.

Figure 3.2 illustrates a basic system architecture for over-the-air management of NFC-based car keys: The secure element contains an application that stores access

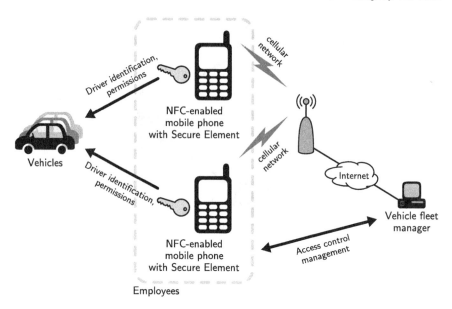

Fig. 3.2 Basic system architecture for over-the-air management of NFC-based car keys (based on [14])

control tokens for one or more cars. The vehicle fleet manager can add, update and revoke these permissions across the cellular network. Every driver can be assigned permissions for a specific vehicle and a specific time span. Lost or stolen car keys (i.e. lost or stolen mobile phones) can be revoked over the cellular network.

Such a system has several benefits: Time-consuming activities like the return/exchange of physical keys are no longer necessary. Moreover, the overall security is increased as users (e.g. employees) can only access their assigned vehicles during a predefined period of time. Furthermore, a fleet management system like this can easily be combined with the automation of other tasks like, for instance, keeping a driver's log.

Nevertheless, using the mobile phone as a car key also has some disadvantages. A mobile phone is not as robust as a regular car key. For instance, the mobile phone may get damaged from external conditions like heavy rain or snowfall. Even worse if the user wears gloves or has both hands full of bags the phone may fall on the ground and break.

3.1.2 Transmission of Data Generated by Vehicle Sensors

NFC can be used to quickly establish other wireless communication channels like Bluetooth or Wi-Fi [11]. Parameters and shared secrets for the wireless link are negotiated across the NFC interface in order to establish the communication channel (out-of-band pairing).

In an automotive environment, a mobile phone can use this functionality to establish a Bluetooth or Wi-Fi link to the computer system of an automobile. This provides a means to connect a vehicle to global networks (i.e. the cellular network and, consequently, the Internet) through the mobile phone.

An application of this connection to the Internet is intelligent systems for traffic optimization. Position and motion information can be transmitted to an online system. This allows for early detection of traffic jams and for energy-efficient guidance of traffic by giving useful tips about optimized speed in relation to green phases of traffic lights. In addition, the collected data could be used to analyze the causes of accidents and to even avoid looming accidents by early detection of critical situations.

In case of a breakdown or a looming breakdown it is possible to find a suitable repair shop and—if necessary—a towing service based on the error diagnostics (e.g. sensor data), the state of the car, the current position and the planned route. For instance, it would be possible to continuously monitor sensors of the motor control unit. As an example, faulty operating conditions of the engine of a car could be detected by analyzing the exhaust fumes based on data retrieved from the lambda probe. Similarly, an increased use of fuel could also be the result of an imminent defect of the engine.

An online monitoring system would allow early detection and—ideally—immediate corrective actions in case of such fault conditions. Otherwise, such problems are possibly detected only after thousands of kilometers (e.g. during scheduled service intervals or after an actual breakdown). Additionally, data on phases of green traffic lights, on traffic jams, and on the current traffic situation in general could be used to adapt parameters of a vehicle (cf. BMW's project Ilena [6]).

While those services could significantly improve energy and economic efficiency as well as the comfort of the driver, they also increase the vulnerability to security threats and attacks: the mobile phone acts as an interface between the automotive computer system, the cellular network and—consequently—the Internet. Moreover, sensitive data is exchanged across the cellular network. On the one hand, this connection is a central part of many approaches to increase the efficiency of automobiles. On the other hand, a connection between the automotive computer system and its surrounding world opens up an avenue for intrusions into that system.

Recent research (cf. [8, 10, 12, 16]) reveals, however, that automotive electronics is sensitive to various attacks that allow a potential attacker to take over control of parts of the (or even the whole) automotive system. Threats range from showing text on a car radio display to putting brakes out of operation or even forcing a car to accelerate [1, 10]. Protection from security issues is, therefore, a core requirement for integrating efficiency-increasing measures based on mobile phones and NFC into vehicles.

These systems could benefit from NFC secure element technology that provides secure storage, encryption and digital signature. Remote systems can use digital signatures to verify the integrity and authenticity of transmitted data. Signature—through the secure element—could even link data to a specific driver. Similarly, encryption assures that only the designated recipient can use encrypted data. Keys

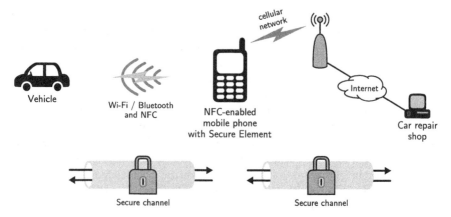

Fig. 3.3 Schematic system architecture for secure data transfer between an automotive computer system and a car repair shop back-end system or a traffic optimization system (based on [14])

for digital signature and encryption can be securely stored inside the secure element. Thus, an attacker can neither extract nor exchange these secret keys in order to intercept communication.

Figure 3.3 shows a schematic system architecture for secure communication between a car and a repair shop back-end system (or an intelligent traffic optimization system): First, the out-of-band pairing functionality of NFC is used to establish a Bluetooth or Wi-Fi link between the car and the mobile phone. Then, the mobile phone secures the communication between the car and the remote back-end system using the secure element. The secure element encrypts every message with the back-end public key and signs each message with the driver's secret key. Messages from the back-end are encrypted with the driver's public key and can, therefore, only be decrypted by that driver's secure element. An additional signature assures the authenticity and integrity of the message. The communication between the mobile phone and the car is protected in a similar manner. Consequently, only the legitimate driver can decrypt messages from the car and can modify the car parameters.

3.1.3 Intelligent Cloud-Based Multimedia Applications

Another group of services that could benefit from integrating NFC into automobiles (in terms of mobility and comfort) are multimedia applications.

An example use-case is a music library in the cloud. Music retailers could offer this service for their customers: Users have their own virtual music libraries that are accessible over the Internet. The libraries are directly linked to an online store where new music can be bought and added to the libraries. The car radio (in-car multimedia system) can be linked to such a library. The radio can then access the media library over the Internet. As with the transmission of sensor data, an NFC-enabled mobile

3.1 Improving Efficiency in Automotive Environments

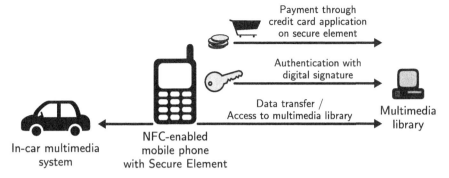

Fig. 3.4 Scenario for access to a cloud-based multimedia service

phone could function as the gateway between the in-car multimedia system and the Internet. Similarly, the secure element could be used to secure the system against malicious activities.

Figure 3.4 depicts a scenario for access to a cloud-based multimedia service. After a connection between the mobile phone and the car has been established using the out-of-band pairing capabilities of NFC, the user is authenticated to the online service based on a digital signature issued by the secure element. After successful authentication, the in-car multimedia system can access the music library. Furthermore, the in-car multimedia system could enable the user to directly buy new music in the online store. In that case, payment could be securely authorized through a credit card stored on the secure element inside the mobile phone.

3.2 Generalized Use-Cases

These use-cases take advantage of several aspects of NFC technology. In order to analyze the security requirements and the current state of security of these features of NFC, it is necessary to derive more general scenarios from these use-cases. Two aspects of NFC-enabled mobile phones have a particular focus throughout all of the above use-cases: the out-of-band pairing capability and the secure element.

3.2.1 Out-of-Band Pairing with NFC

The NFC link itself only supports data rates of up to 424 kbps. While this is sufficient for transferring small amounts of data, it is too slow for applications that need to exchange a large data volume. There exist other, more suitable communication technologies—for instance Bluetooth and Wi-Fi—for the exchange of large volumes of data. However, configuration of these fast wireless carriers typically requires complex configuration procedures or even the knowledge of device names, addresses and parameters [11].

To maintain the simplicity of establishing connections and exchanging information with a simple touch gesture through NFC, while at the same time being able to achieve high data rates, NFC can be used as an "out-of-band" channel for establishing other wireless communication technologies between two devices ("pairing"). This capability is, therefore, called *out-of-band pairing*. The NFC interface is used to automatically negotiate available interfaces, addresses, parameters and shared secrets in order to quickly and easily establish a fast wireless communication channel for the actual data exchange.

The NFC Forum created the *Connection Handover* technical specification [13] which defines a reference application for negotiating alternative communication channels. The reference application consists of NDEF record definitions and handshake protocols. Two connection handover protocols are available: negotiated handover and static handover.

Negotiated handover is intended for use with NFC peer-to-peer mode. With this protocol, two NFC devices can find a common alternative carrier that best suits both devices. The two devices exchange lists of their supported communication interfaces and choose the most efficient interface that both have in common. Besides the type of the interface, they also exchange interface parameters, each other's communication endpoint addresses and can even establish a shared secret for securing the communication.

Static handover is intended for use with NFC tags and NFC reader/writer mode. With this protocol an NFC device can retrieve static information about available communication interfaces, interface parameters and communication endpoint addresses that are stored on NFC tags. Thus, this handover protocol enables simple out-of-band pairing even for non-NFC devices.

3.2.2 Secure Element

The secure element serves the third mode of operation of NFC: card emulation mode. In this mode, an NFC device acts as a contactless smartcard and can interface with existing RFID reader infrastructure.

Typically, secure elements are standard smartcard microchips as used for contact and contactless smartcards. Instead of (or in addition to) the classical contact-based or contactless interfaces, a secure element chip is equipped with a dedicated interface to connect to the NFC controller. The NFC controller can use a secure element in two modes: external mode and internal mode. In external mode, the NFC controller ties the secure element to the NFC antenna and, therefore, provides the RF (radio frequency) interface to interact with RFID readers. In internal mode, the NFC controller links the secure element to the application processor. Thus, applications that run on the application processor can access the secure element as if it was a smartcard attached to a reader device.

3.2 Generalized Use-Cases

The use-cases in Sect. 3.1 perform various tasks with the secure element:

- The secure element is used as a security device for identification, authentication and messaging security (i.e. for secure storage, digital signature and encryption).
- The secure element is used as a digital wallet to securely store payment cards.
- Data and applications contained within the secure element are managed over the mobile phone network ("over-the-air").

3.2.2.1 Security Device

A secure element is based on the same hardware and software platforms as regular smartcards. Consequently, it provides the same set of functionality and protection as regular smartcards. A secure element contains secure storage, provides a secure execution environment and has hardware-based support for cryptographic operations. The IC (integrated circuit) itself, as well as its software, is protected against various attacks that aim for retrieval or manipulation of stored data and processed operations.

A secure element can store secret cryptographic keys in a way that they cannot be read from outside the secure element. Thus, it can operate as an uncloneable security device that can perform decryption and digital signature based on such keys. Similarly, public keys can be stored in a way that only authorized parties can update them. Consequently, a secure element can operate as an unmodifiable security device for encryption and verification of digital signatures.

In external mode, this capability can be used to identify and authenticate the holder of an NFC device to an RFID smartcard reader over the contactless RF interface. Cryptographic key agreement and authentication protocols assure mutual authentication between the secure element and the external device and also establish shared secrets for guaranteed end-to-end encryption for secure exchange of user information and credentials.

The security device capability is not only useful for external mode. In internal mode, the secure element could be used to securely authenticate to web/cloud services just in the same way as it does with external reader devices. The secure element could also be used by apps on the application processor to encrypt and decrypt messages based on securely stored keys.

3.2.2.2 Digital Wallet

A digital wallet is a specialized security device dedicated to payment cards, loyalty cards, digital money, and coupons. Hence, the security device capability of a secure element fulfills all the security requirements of a digital wallet.

For example, a credit card requires unmodifiable storage of card identification data (e.g. credit card number, validity period, card holder information). It also requires uncloneable storage of secret keys and the capability to perform cryptographic operations for authentication and authorization of payments.

Virtual credit cards (or payment cards in general) stored inside the secure element can be useful in both external mode and internal mode. In external mode, an NFC device can be used just like any other contactless credit card with any contactless-enabled credit card terminal. In internal mode, the virtual credit card could be used for secure "card-present" payment in apps and on web pages.

3.2.2.3 Over-the-Air Management

While classical smartcards are typically only configured once *before* they are distributed to their users, this approach is not feasible with secure elements. The set of services that a secure element is used for can change at any time while the secure element is in the field. Therefore, it must be possible to manage the applications and data stored on the secure element over the cellular network or the Internet. Typically, the secure element as a whole is managed by a trusted service manager (cf. [7]). This trusted service manager can install, configure, disable and uninstall applications and can lock lost or stolen secure elements as a whole. Individual applications could also be managed directly by their respective service providers.

For most types of secure elements, the operating system or an application running on the application processor need to tunnel the communication between the secure element and the remote managing entity through the internal mode of the secure element. Only for a universal integrated circuit card (UICC) based secure element, the management link could be established directly through the subscriber identity module (SIM) application and does not necessarily need to pass through the application processor. In any of these cases, application and data management can be performed over an encrypted and mutually authenticated communication channel. Security and management protocols are defined by GlobalPlatform (cf. [5]).

3.3 Identification of Security Aspects

The various application scenarios have different requirements with regards to security. For further analysis, the three operating modes of NFC devices are examined separately.

3.3.1 Peer-to-Peer Mode

Negotiated connection handover and peer-to-peer communication in general require authenticity and integrity protection because recipients want to be sure that they receive the data that they expect to receive and that no attacker is able to present forged data to them. In addition, some peer-to-peer communication (in particular the exchange of shared secrets and other sensitive data) also requires confidentiality.

Otherwise, an attacker could acquire sensitive data by tapping into the communication. If shared secrets for encryption of an alternative wireless carrier are intercepted by an attacker, the attacker could then also eavesdrop on the communication on that carrier. Without proper authenticity and integrity protection, an attacker could possibly even manipulate the communication. In the case of a negotiated connection handover, this could allow an attacker to manipulate endpoint addresses for the alternative wireless carrier. Thus, attackers could redirect the established wireless link through their own equipment and, consequently, take full control over that communication.

However, there exist protocols to prevent such attacks. For instance, the NFC-SEC protocol suite provides mechanisms for key agreement, encryption and integrity protection of the data exchange protocol across the NFC peer-to-peer communication channel. The protocol suite has been first standardized in ECMA-385 [2] and ECMA-386 [3] in 2008 and has been adopted by ISO in the ISO/IEC 13157 series [9] in 2010.

3.3.2 Reader/Writer Mode

Static connection handover has similar requirements as negotiated handover. Recipients want to be sure that they receive the data that they expect to receive and that no attacker is able to present forged data to them. As a result, static connection handover needs authenticity and integrity protection. As opposed to negotiated handover, NFC tags with static handover information are typically located in public places and anybody should be able to read them. Thus, confidentiality is usually not necessary.

Reading tags in general demands a high level of trust, authenticity and integrity protection. On the one hand, NFC tags are usually located in publicly accessible places. Therefore, it could be fairly easy for an attacker to mess with these tags. On the other hand, however, users expect that it is safe to tap these tags with their NFC-enabled mobile phones. At the same time, they expect that a tag triggers an immediate action on their device (e.g. establish a wireless connection, open a web page, send an SMS message). This requires a method to establish trust in the contents of NFC tags in order to create a good user experience for "tagging" applications (i.e. applications that involve interaction with NFC tags).

3.3.3 Card Emulation Mode

The use-cases identified in this chapter require the secure element to provide secure storage memory and cryptographic operations for authentication, message encryption and authorization of payment transactions. In general, the secure element, therefore, needs to guarantee authenticity, integrity protection, confidentiality and non-repudiation.

Card emulation mode has a significant advantage in terms of security: It is tightly linked to the security of regular smartcards. Regular smartcards have been used for security critical tasks for a long time. Most modern smartcard microchips are certified according to Common Criteria security standards for smartcards. The chips are designed to be tamper-proof and robust against many physical and electrical attacks. Smartcard operating systems are also carefully designed to mitigate all kinds of attacks. Smartcard application software has existed for a long time too. Thus, application software, protocols, and standards had a long time to mature.

A secure element is just a regular smartcard chip with specialized interfaces to its surrounding world. As a consequence, it inherits all the security properties of regular smartcards. Nevertheless, card emulation mode introduces a new path to the secure element beyond the external contactless interface. To fulfill the use-cases in this chapter, many applications on the secure element would be accessible from the application processor through internal mode. While communication in external mode is only possible when the NFC device is in read range of a contactless reader device, applications on the application processor may access the secure element at any time. Moreover, applications running on the application processor also have the ability to use the cellular network connectivity. While this is, for instance, necessary for over-the-air management, it might open a new path for intruders.

References

1. carIT: Autos brauchen Schutz vor Hacker-Attacken. carIT Mobilität 3.0. http://www.car-it.com/secunet-infotainment-hacker-sicherheit/id-0012184 (2010)
2. Ecma International: ECMA-385: NFC-SEC: NFCIP-1 Security Services and Protocol (2008)
3. Ecma International: ECMA-386: NFC-SEC-01: NFC-SEC cryptography standard using ECDH and AES (2008)
4. European Commission: Progress Report on the Single European Electronic Communications Market (15th Report). Commission Staff Working Document SEC (2010) 630 final (2010)
5. GlobalPlatform: Card specification. Version 2.2.1 (2011)
6. Grünweg, T.: Lernfähiger Routenplaner: Big Navi is watching you. http://www.spiegel.de/auto/aktuell/0,1518,608617,00.html. Spiegel Online (2009)
7. GSMA: Mobile NFC services. Version 1.0. White paper (2007)
8. Hoppe, T., Kiltz, S., Dittmann, J.: Security Threats to Automotive CAN Networks—Practical Examples and Selected Short-Term Countermeasures. In: Computer Safety, Reliability, and Security. LNCS, vol. 5219/2008, pp. 235–248. Springer, Berlin Heidelberg (2008). doi:10.1007/978-3-540-87698-4_21
9. International Organization for Standardization: ISO/IEC 13157: Information technology—Telecommunications and information exchange between systems—NFC Security (Parts 1–2) (2010)
10. Koscher, K., Czeskis, A., Roesner, F., Patel, S., Kohno, T., Checkoway, S., McCoy, D., Kantor, B., Anderson, D., Shacham, H., Savage, S.: Experimental security analysis of a modern automobile. In: Proceedings of the IEEE Symposium on Security and Privacy (S&P), pp. 447–462. IEEE, Oakland, CA, USA (2010). doi:10.1109/SP.2010.34
11. Langer, J., Roland, M.: Anwendungen und Technik von Near Field Communication (NFC). Springer, Heidelberg (2010)

12. Larson, U.E., Nilsson, D.K.: Securing vehicles against cyber attacks. In: Proceedings of the 4th Annual Workshop on Cyber Security and Information Intelligence Research (CSIIRW '08). ACM, Oak Ridge, TN, USA (2008). doi:10.1145/1413140.1413174
13. NFC Forum: Connection Handover. Technical specification, version 1.1 (2008)
14. Roland, M., Langer, J., Bogner, M., Wiesinger, F.: NFC im Automobil: Software bringt Ökonomie und braucht Sicherheit. In: Höfler, L., Kastner, J., Kern, T., Zauner, G. (eds.) Energieeffiziente Mobilität, Informations- und Kommunikationstechnologie, pp. 112–119. Shaker, Aachen (2010)
15. Steffen, R., Preißinger, J., Schöllermann, T., Müller, A., Schnabel, I.: Near Field Communication (NFC) in an automotive environment. In: Proceedings of the Second International Workshop on Near Field Communication (NFC 2010), pp. 15–20. IEEE, Monaco (2010). doi:10.1109/NFC.2010.11
16. Wolf, M., Weimerskirch, A., Wollinger, T.: State of the art: embedding security in vehicles. EURASIP J. Embed. Syst. **2007**: 074706 (2007). doi:10.1155/2007/74706

Chapter 4
Related Work

There have been several research activities focused on the security and privacy of Near Field Communication (NFC) and its underlying Radio Frequency Identification (RFID) technologies during the last couple of years. To assess the current status of NFC security and privacy, it is necessary to collect preceding research results and to analyze the issues and solutions identified in them.

4.1 Communication Protocol

Haselsteiner and Breitfuß [45] thoroughly evaluate the security and privacy aspects of the NFC signaling layer (as defined in ISO/IEC 18092 [51]). They conclude that the most serious threats are eavesdropping and data corruption. Eavesdropping is an attack where the attacker listens to the RF (radio frequency) waves of the communication between two NFC devices from a longer distance. Data corruption is a scenario where the attacker sends arbitrary data on the same frequency as the real sender, so that the actual receiver is unable to decode the real sender's data, effectively resulting in a denial-of-service. Attacks which they identified as also possible (but difficult) are data modification and data insertion. Data modification is a scenario where an attacker changes specific bits in the communication to selectively manipulate the meaning of the data. However, Haselsteiner and Breitfuß state that data modification is only possible for certain modulation and coding schemes. In a data insertion attack, the attacker uses the idle time between a command and a response to send an alternative response. Another attack scenario identified by them is the man-in-the-middle attack, where the attacker routes all communication between the two NFC devices over a third device. This gives the attacker full control over the data exchange. However, Haselsteiner and Breitfuß conclude that man-in-the-middle attacks are impossible due to the proximity between the two NFC devices.

As a universal solution against eavesdropping, data modification and data insertion, they suggest to use a secure channel that provides encryption and integrity protection [45]. This is also the approach that has been followed in recent standardization activities: ISO dedicated the ISO/IEC 13157 [52] series to NFC security related standards. First standards of this series define a secure channel for NFC and have been adopted by ISO in early 2010. However, it is unknown if there are already applications based on these standards. Also, these standards have not yet been adopted into NFC Forum specifications. As an alternative to the security protocols defined in ISO/IEC 13157, the TLS Working Group of the Internet Engineering Task Force (IETF) published a draft [93] on using transport layer security (TLS) over the NFC Logical Link Control Protocol (LLCP) layer to create authenticated and encrypted LLCP connections.

4.2 Flaws in Legacy Contactless Chip Card Systems

Besides the NFC-specific peer-to-peer communication mode, NFC devices are capable of interacting with proximity RFID chip card systems in reader/writer mode as well as in card emulation mode. NXP MIFARE Classic is one of these legacy chip card systems that can be read and written by many NFC-enabled handsets, and that can be emulated by many secure elements. MIFARE Classic, a technology that celebrated its 20th anniversary in 2014, is still in wide use. Therefore, its integration in NFC-enabled mobile phones—and the resulting backward compatibility to existing contactless infrastructures—is considered essential for a success of NFC.

Nevertheless, the cryptography of MIFARE Classic has been broken in 2007. Nohl et al. [79, 80] reverse-engineered the cryptographic circuit of a MIFARE Classic card and obtained the secret MIFARE *Crypto-1* cipher. They evaluated the cipher with regard to its weaknesses. They identified several weak parts (a linear feedback shift register (LFSR) based random number generator that derives its value from time, a non-linear component in the feedback loop, and an output derived from a fixed subset of bits) and conclude that the security of the proprietary Crypto-1 algorithm is primarily based on obscurity. Nohl [78] performed further cryptanalysis of Crypto-1 to improve the performance of attacks.

Independently, de Koning Gans et al. [62] found a method to recover the key stream of recorded MIFARE Classic transactions. They used this information to read parts of a MIFARE Classic card without actually knowing the secret key.

Courtois et al. [16] propose an algebraic method to recover the secret key of the Crypto-1 cipher from a known initialization vector and 50 bits of the encrypted output. They claim that their attack can be performed within several hundred seconds on a standard personal computer (PC) platform.

By reverse-engineering the communication between a MIFARE reader and a MIFARE Classic card, Garcia et al. [32] found two methods to attack a MIFARE Classic system: With the first method, an attacker can obtain the secret keys from a reader without possession of an actual card. The second method allows obtaining the keys as well as the plaintext of the communication from a recorded communication between a reader and a card.

4.2 Flaws in Legacy Contactless Chip Card Systems

Teepe [92] summarizes the vulnerabilities of MIFARE Classic and also presents strategies to reduce their threats.

Further analysis by Garcia et al. [35] revealed that MIFARE Classic cards can be attacked even without possession of a genuine reader device (i.e. a reader that knows the secret keys). Instead, weaknesses of the authentication protocol and the cipher can be abused to compute the secret keys with only a few (unsuccessful) authentication attempts and pre-calculated lookup tables. Courtois [15] proposes an improved method of the card-only attack that does not require huge lookup tables and only requires a few thousand authentication attempts. As a consequence, MIFARE Classic is considered to be completely broken.

Besides MIFARE Classic, other proprietary contactless chip card systems become a target for attacks too. An example is HID iClass which has been dismantled by Garcia et al. [33, 34].

While many broken products like MIFARE Classic have been superseded[1] by more robust products (e.g. MIFARE DESFire EV1), compatibility of NFC to these legacy systems has its downside: It potentially leads to the believe that NFC as a whole is insecure and even dangerous because it is compatible to insecure technology.

4.3 Attacks on Contactless Smartcards

Besides potential vulnerabilities as a result of protecting cryptographic algorithms in proprietary RFID systems by obscurity instead of robustness, there are also other attack scenarios on contactless smartcards. For instance, power analysis (a form of side channel analysis) could be used to reveal information about the operations performed on a microchip (cf. Kocher et al. [61]). Thus, power analysis has the potential to retrieve secret information that is processed in cryptologically robust cryptographic operations (e.g. secret keys). A system that has proven vulnerable to this class of attacks is MIFARE DESFire. This has been shown by Oswald and Paar [81].

Typically, modern smartcard microchips are protected against various kinds of physical and logical attacks by sensors and counter measures in hardware, and by a robust design process (cf. [67]). Chip hardware and software, as well as their design processes, are usually evaluated and certified according to high security standards. Examples for such evaluation criteria are the various smartcard-related Common Criteria protection profiles (cf. [24, 25] for examples of integrated circuit (IC) hardware related protection profiles and [90, 91] for examples of card operating system related protection profiles). While this thorough design process mitigates most known attack scenarios, new and—up to now—unconsidered attack scenarios are still possible, as has been demonstrated by Barbu et al. [6].

---

[1] Even though MIFARE Classic has a more robust successor for several years, it is still widely in use.

Apart from new vulnerabilities, there is also a class of well-known attack scenarios on (contactless) smartcards: relay attacks. Described by Conway [14] as the "Grandmaster Chess Attack" and by Desmedt et al. [18] as the "mafia fraud", relay attacks are a means for impersonating someone else. Desmedt et al. [18] describe their mafia fraud attack as follows:

> A identifies himself to B. [B] is collaborating with C and C impersonates A [...] Then, D checks the identity of C who is claiming to be A. [...] A and D are not aware of the [...] fraud. [...] While D is checking the identity, C and B [...] sit in the middle between A and D. So B and C pass all questions and all answers related to the [...] identification going from D to A and vice-versa.

Thus, the relay attack (*mafia fraud*) can be seen as a simple extension of the communication channel between A and D. Ideally, A and D do not notice any difference between the relay scenario and direct interaction.

A variation of the relay attack is the wormhole attack in wireless networks. It has been introduced by Hu et al. [49, 50]. The wormhole attack does not only tunnel the communication between two endpoints A and D but instead relays packets received from one or more senders at one location to one or more receivers at another location. Hu et al. [49] describe the wormhole attack as an attack, where "[...] an attacker receives packets at one point in the network, 'tunnels' them to another point in the network, and then replays them into the network from that point." They introduce "packet leashes" as measure to detect wormhole attacks. A packet leash is additional authenticated location and time information that is attached to each packet. Based on that information, the recipient can then determine if the packet traveled across the expected path or across a wormhole.

In 2005, Hancke [38] first applied relay attacks to ISO/IEC 14443 Type A based smartcard systems. His relay system consists of a "mole" (relay reader; B in the mafia fraud scenario) and a "proxy" (relay card emulator; C in the mafia fraud scenario). The mole and the proxy forward the demodulated and decoded bits of the data link layer communication of the real card and the real reader through a fast ultra high frequency (UHF) channel. His test setup could successfully relay communication over a distance of up to 50 m. Though, he admitted that there would be timing issues during the anti-collision phase if there were multiple cards in the field of the real reader. Hancke [38] explains that "this attack is invisible to application layer security [...]" Thus, cryptographic protocols for confidentiality, authentication and integrity cannot prevent relay attacks. Hancke [38] concludes that "if a contactless card could be read while in a pocket, purse or wallet, a thief might be able to engage in the act of digital pickpocketing while standing next to or merely walking past his victim." The victim, therefore, would not even notice the attack.

Kfir and Wool [56] describe a similar system. Through additional amplification and filtering in the relay reader, they were able to access a victim's card from a distance of up to 50 cm, which is a significant improvement over typical reading distances of proximity RFID systems.

Drimer and Murdoch [20] show that relay attacks are also possible with contact smartcards. Their scenario relays EMV *Chip & PIN* credit card transactions.

4.3 Attacks on Contactless Smartcards

Hancke [38], Kfir and Wool [56] and Hancke et al. [44] propose several countermeasures against relay attacks:

- The RF interface of the card, when not in use, could be physically shielded with a Faraday cage (e.g. aluminum foil).
- The card could contain additional circuitry for physical activation and deactivation (e.g. an on-off switch).
- In addition to the card, a secondary authentication factor (e.g. password, PIN, biometrics) could be used to verify legitimate use of the card.
- Hancke [39] states that "an attacker executing a relay attack cannot avoid causing a delay in the system." This suggests relay attacks can be prevented by determining the distance through these delays. Thus, timing constraints could be used to limit the delay that may be introduced by the relay channel. However, Hancke et al. [44] conclude that the timing constraints of ISO/IEC 14443 are too loose to provide adequate protection against relay attacks.
- Distance bounding protocols can be used on fast channels to determine the actual distance between the card and the reader.

Distance-bounding is a sophisticated and effective countermeasure. This method allows determining the distance between the real card and the real reader. They rely on the fact that the "maximum propagation speed of the communication medium is constant and known" [40]. Distance-bounding protocols for RFID and NFC have been designed and evaluated in many publications (cf. [5, 40–42, 59, 60, 83]). However, conventional channels and, in particular, the ISO/IEC 14443 communication channel are too slow for accurate distance bounding [43].

Besides distance-bounding protocols there exist also other approaches to counter relay attacks at the communication channel. For instance, Choudary and Stajano [12] propose a protocol based on inducing noise on the communication channel. Though, they admit that their scenario is idealized and would not withstand a real-world attack without further research.

4.4 Security and Privacy Aspects of NFC Devices

Several threats to security and privacy of NFC devices have been identified by Madlmayr et al. [69]:

- relay attacks,
- skimming of applications on the secure element,
- access to the secure element from the host controller,
- unencrypted peer-to-peer communication,
- privacy issues and identity spoofing due to static unique identifiers, and
- phishing and denial-of-service through manipulation of NFC tags.

Access to the secure element from the host controller and phishing/denial-of-service through manipulation of NFC tags are the basis for the research conducted in this thesis.

4.4.1 Tagging and Peer-to-Peer Communication

Madlmayr et al. [69] assume that "the inhibition threshold of touching a tag or a reader with the mobile phone is probably much lower than making an intended connection with a wire." Therefore, the average user will not be able to distinguish forged tags from genuine tags. This makes applications based on NFC Data Exchange Format (NDEF) tags potentially vulnerable to data modification. By modifying tag contents or by replacing NFC tags, an attacker could perform various types of phishing and social engineering attacks.

In 2008, Mulliner [73] evaluated various attack scenarios on existing NFC-enabled mobile phones through their reader/writer mode capabilities (cf. *tagging*). He identifies several suitable attack targets. Among them are bugs in the mobile phone system, bugs and design flaws in mobile phone applications and the tag infrastructure itself.

Mulliner [73] explains that there are several possibilities to spoof tags. Some tags that are deployed in the field are not write-protected at all. In that case attackers can simply overwrite the tags with their spoofed NDEF data. Some tags are protected by weak write keys. Particularly with MIFARE Classic, which is often used as an alternative to standard NFC tags, write keys can easily be discovered (cf. Sect. 4.2). Mulliner [73] has a solution even if the tags are permanently write-protected: An attacker could destroy or shield the original tag and stick a new tag on top of it.

Mulliner's analysis [74] of the Nokia 6131 NFC mobile phone reveals several flaws in the NDEF implementation and the web browser that make the phone susceptible to content spoofing with NDEF smart poster records. All of these spoofing attacks follow a similar pattern: The phone usually displays the title record followed by the uniform resource identifier (URI) record (Internet address, telephone number, etc.). Consequently, an attacker could use a specially crafted title record to show a falsified URI and push the real target URI off the screen. A user is likely to fall for this trick without even noticing the manipulated URI. Therefore, attackers could take advantage of this approach by redirecting the users to phishing websites or by redirecting telephone calls or SMS messages to their own premium rate service [74]. As most problematic he identified that JAR files containing executable application code are automatically downloaded to the device when referenced through a URI [74]. The user is then only one click (a confirmation dialog to run the application) away from executing the application.

Verdult and Kooman [95] found several vulnerabilities in the Nokia 6131 NFC and Nokia 6212 mobile phones that allow an attacker to initiate Bluetooth connections and Bluetooth file transfers with specially crafted NFC tags and peer-to-peer communication. With one attack scenario an attacker can install an application on the victim's phone over an OBEX (OBject EXchange) file transfer connection without requiring user consent. Another attack scenario establishes a pairing between the phone and another Bluetooth device with presenting "only one vague notification" [95] to the user. They further describe how the two attack scenarios can be combined to establish a Bluetooth connection to the Nokia PC Suite interface of the phone that can be used to gain full write access to the phone memory. An attacker can abuse that write access to install an application and to elevate the privileges of that application to a level where it is allowed to access sensitive information and functionality of the phone.

Mulliner [74] employed black-box testing (*fuzzing*) to find vulnerabilities in the implementation of the NDEF parser of Nokia's first NFC-enabled mobile phones. He used NFC tags with malformed NDEF messages to trigger invalid input conditions of the NDEF parser. He found several input conditions that lead to errors and even crashes of the mobile phones. Attackers could potentially abuse such conditions to trigger code execution exploits. Though, Mulliner admits that no code injection techniques existed for that platform at the time of his writing. Despite the fact that such issues were known for quite some time, Mulliner's recent research [75] revealed that similar problems still existed on the Nexus S, Google's first NFC phone.

In 2012, Mulliner [76] refined his black-box testing for Android NFC phones. He used binary instrumentation to add logging and data manipulation functions to the Android NFC stack. As a result, he could sniff the communication between the NFC stack and the NFC controller hardware. Moreover, he was able to simulate NFC tags with his setup to perform automated fuzzing without the need to touch a physical tag.

Yet another approach was used by Miller [72]: He performed fuzzing with a card emulator device that repeatedly emulates NFC tags. This method also has the advantage that there is no need to touch a physical tag, but without the need to manipulate the software on the mobile phone. Miller found several conditions that trigger crashes in native code that possibly result into memory corruption vulnerabilities.

Miller [72] also found that, starting with Android 4.0, Internet addresses received from NFC tags or over *Android Beam* (the peer-to-peer mode implementation of Android) are automatically opened in the web browser. Similarly, other applications can register for specific NDEF data types and URIs too. Miller [72] concludes that "[...] if an attacker can get the device to process an NFC tag, they can get it to visit a web site of their choosing in the [web] browser with no user interaction." As a consequence, the potential attack surface of the web browser is now open to NFC. Mulliner [75] as well as Benninger and Sobell [9] found that this automatic processing without the need for user confirmation can be used to mount phishing attacks and to trigger unintended actions in the context of the user (e.g. to trigger a check-in on Foursquare).

4.4.2 Protection for Tagging and Peer-to-Peer Communication

Madlmayr et al. [69] suggest digital signatures as a suitable countermeasure against tag spoofing. When NDEF messages are protected by digital signatures, an NFC device can verify the authenticity and integrity of received NDEF records. Schoo and Paolucci [88] follow a different approach: They propose that tag spoofing can be reliably prevented by registering all genuine tags in a database back-end and by using a certified application on the NFC device that compares the tag data with the data stored in that back-end database. Wu et al. [96] follow a similar approach:

Tags only contain a unique identifier that is dynamically mapped to content (either a link to a website or a user poll) in a back-end database. Their system also allows authentication of users and access control to restrict access to tag content.

Digital signatures are the solution that has been chosen by the NFC Forum. Kilås [58] evaluated several digital signature algorithms regarding their feasibility and performance on mobile Java platforms. He also created a reference implementation [57] for Java ME (Java Platform, Micro Edition). The NFC Forum released the first draft of their Signature Record Type specification in early 2010 [13].

Rosati and Zaverucha [84, 85] conclude that this Signature Record Type specification is impractical due to its huge space requirements for storing the signature and the certificate chain on a tag. They explain that the size of the signature record would be much greater than that of the NDEF messages typically signed with it. They note that these space requirements would exceed the memory sizes of most of the available NFC tag types. This is especially true for low-cost tags. To overcome this issue, Rosati and Zaverucha [85] propose the use of ECQV (Elliptic Curve Qu-Vanstone) implicit certificates together with ECDSA (Elliptic Curve Digital Signature Algorithm). This would drastically reduce the size of the signature record (factor of 4–9).

Rosati and Zaverucha [84, 85] further clarify that the Signature Record Type specification lacks a definition of the public-key infrastructure (PKI). A PKI defines a set of the rules for certification of signature keys and for trust in signatures. However, they conclude that these definitions are necessary for signatures to be useful and to achieve an improved user experience in tagging use-cases.

4.4.3 Integration of Secure Elements into Mobile Phones

The integration of secure elements into mobile phones is often seen as a great advantage to improve usability and security of the mobile phone, the secure element and its applications. Anderson [2] states the following:

> This technology holds out the prospect of solving the problem of a trustworthy user interface. The plan is that instead of being a relatively dumb device, your credit card will be an application on your mobile phone. You bring your mobile into close proximity with the merchant terminal, an application displays the sale amount, you authorise this, and the transaction goes through.

In [37], the mobile handset is proposed to be used as a secure display and as a secure PIN entry device for mobile contactless payment. This concept has been adopted in the EMV specifications: [23] allows "Consumer Device CVM", a cardholder verification method where the PIN is entered on the device that is used as the contactless payment token. This approach is also in line with Drimer and Murdoch [20] who argue that customers cannot easily distinguish forged from genuine credit card terminals.

Besides using the NFC-enabled mobile handset as a trusted user interface for mobile contactless payment, the secure element could also be used for secure Internet purchases through the mobile phone web browser. Attard [4] proposes a card-present payment scheme for in-browser credit card payment on NFC-enabled mobile phones.

4.4 Security and Privacy Aspects of NFC Devices

While this scheme uses NFC to interface an external credit card, it could also be used in combination with a credit card stored in the secure element.

Another benefit of the secure element in comparison to a standalone smartcard is over-the-air management. Applications on the secure element can be installed, controlled and removed throughout its whole lifecycle. If, for example, a device is lost or stolen, the secure element and all its applications can be wiped through the over-the-air management link.

Besides bringing new advantages over regular smartcards, the secure element is also supposed to add current security properties of regular smartcards (secure data storage, secure execution environment and hardware-based cryptography) to the mobile phone. Thus, it enables apps on the mobile phone to benefit from these features.

However, combining the mobile phone and the secure element does not only bring advantages. Madlmayr et al. [69] state—with regard to the interaction between the application processor ("host controller") and the secure element—that "applications running on the host controller need to authenticate against the secure element before a communication can be established." They suggest to "[...] implement a certificate based authentication between the application running on the host [controller] and the applets in the secure element."

A guideline for restricting access to the secure element is proposed in [89]:

> A good practice is to require all phone applications that need to communicate to the secure element to be authenticated by a trusted entity [...] The phone's operating system will then prevent access to the secure element APIs by any non-trusted applications.

This implies a trust relationship between the mobile device application processor (including the operating system that runs on it) and the secure element that usually does not exist on current mobile phone architectures. As a result of this lack of trust, code that runs on the application processor poses an additional threat vector to the secure element. Nevertheless, applications that rely on the communication between the application processor and the secure element exist (e.g. Van Damme et al. [94]).

4.4.4 Mobile Phones as Attack Platforms

Francis et al. [28, 29] demonstrate that NFC-enabled mobile phones are not only a target for attacks, but that they could also be used as a platform to carry out attacks on ISO/IEC 14443 based systems. Their research focuses on the fact that NFC devices can operate in both reader/writer mode and in card emulation mode. Thus, they can be used to read, write and emulate contactless smartcards. As a consequence, an attacker could use such a mobile phone to secretly copy a contactless token (e.g. by bringing the mobile phone close to the victim's pocket). Moreover, the attacker could then transfer (*clone*) that data onto the secure element in the mobile phone. The secure element can then be used in card emulation mode to skim the original token. Benninger [8] suggests a similar approach to clone the access token for an

NFC-enabled lock: Instead of using the secure element, the access control card is cloned onto a blank card by using an NFC-enabled handset. However, Francis et al. [29] admit that such cloning and skimming is only possible if the token does not use strong cryptography to communicate with the reader. In that case it would not be possible to retrieve secret/protected data from the token.

Besides cloning tags and smartcards, it is also possible to use reader/writer mode to carry out data modification attacks on tags and contactless smartcards. An example for such a scenario is an attack on MIFARE Ultralight based transport tickets described by Benninger and Sobell [10]. They found that certain transport systems use tickets that store information about the "rides left on [the] card" in an unprotected memory area. Consequently, the ticket memory can be restored to its initial value after each use. As a result, the ticket can be "refilled" an infinite number of times—effectively allowing unlimited use of the transport system.

Anderson [2] suggests that NFC-enabled mobile phones are an ideal platform for relay attacks. With regard to contactless credit card fraud he explains the following:

> [...] NFC is likely to make middleperson attacks much easier. At present, such an attack requires the construction of custom hardware; in future, an attack could be carried out by software installed on commodity mobile phones. One phone could act as a rogue merchant terminal to the cardholder, and communicate with another [phone] that acts as a card to a merchant elsewhere.

Francis et al. [30] evaluate this possibility to use NFC-enabled mobile phones as a platform for relay attacks. In their proof-of-concept, they relay peer-to-peer communication over longer distances. They successfully relay the communication between two mobile phones at the NFC Data Exchange Protocol (NFC-DEP) layer using two other NFC-enabled mobile phones. They use Bluetooth as the relay channel. As a countermeasure, they propose to integrate location information into NFC transactions (cf. packet leashes as a countermeasure to wormhole attacks [50]). They name GPS (Global Positioning System) and broadcast cell identification as accurate sources of location information.

Francis et al. [31] extend this peer-to-peer relay scenario to ISO/IEC 14443 based smartcard systems. They show that NFC-enabled mobile phones can be used in "soft-SE" mode (card emulation through software on the application processor of the phone). Thus, two mobile phones can be used to relay smartcard communication on the application protocol data unit (APDU) layer by using one as a reader and one as "soft-SE". The two relay devices can then forward the communication over any interface that both devices have in common (typically Bluetooth, Wi-Fi or the cellular network).

4.5 Mobile Phone and Smart Phone Security

When NFC functionality is added to a mobile phone, a critical factor for overall security is the security of the whole device software stack. Already in 2004, Gowdiak [36] revealed several vulnerabilities in Java ME that allow escaping the

4.5 Mobile Phone and Smart Phone Security

restrictions of the Java virtual machine (KVM) sandbox on Java ME devices. Using these vulnerabilities, an attacker could elevate the privileges of an application and access otherwise restricted system functions. An attacker could even access the whole device memory. Gowdiak successfully verified his results on a Nokia 6310i but notes that many other devices are potentially vulnerable.

Lately the topic of mobile phone security experiences significantly increasing awareness. Recent research activities include the assessment of vulnerabilities and threats and the uncovering of actual attack scenarios. An important factor for the rise in malicious activities is the concept of "apps" in modern smart phones. In the past, a mobile phone contained a fixed set of applications and the average user could not easily extend this functionality. Today, an important aspect of smart phones is their extensibility. Every user can easily extend the functionality of their own devices by downloading an application ("app") from an online market place.

Jeon et al. [55] analyzed the vulnerabilities and threats in smart phone security. They identified vulnerabilities caused by implementation errors, incompatibilities, user unawareness, improper configuration, social engineering, loss of smart phones and the interaction of a smart phone with its environment. While connectivity features (Internet, wireless networks, etc.) make smart phones "useful and most popular", these same features open up various paths for intruders [55]. The threats identified by Jeon et al. comprise malware, attacks through (wireless) networks, denial-of-service, break-in attacks, malfunction, phishing, loss of devices and platform alteration.

Kooman [64] discovered a vulnerability in Nokia's proprietary PC Suite interface which can be accessed over Bluetooth. The vulnerability allows modification of the certificate store of Nokia's S40 phones. Kooman created an exploit [63] to install X.509 [54] certificates on these devices and to enable the certificates for code-signing, which is otherwise not possible on S40 devices. Thus, an attacker can use this exploit to elevate the privileges of applications installed on a Nokia S40 phone. Verdult and Kooman [95] use this vulnerability to elevate the privileges of an application into the operator or the manufacturer domain, what gives them full control over security critical system functions.

Davi et al. [17] investigated the security model of the Android platform. The security mechanism of Android consists of a combination of discretionary access control for file system access, sandboxing for application execution, mandatory access control for inter-component communication, component encapsulation and application signing. They state that "Android does not deal with transitive privilege usage, which allows applications to bypass restrictions imposed by their sandboxes." Consequently, an application with lower privileges could potentially access an application with higher privileges in such a way, that it could use these higher privileges for its own purposes.

Armando et al. [3] found a vulnerability of the Android operating system that can be exploited by malicious applications to perform a denial-of-service attack and even force a device to reboot.

According to McAfee Lab's quarterly threat report [71], the trend towards threats and malware for mobile platforms has dramatically increased in 2012. Especially for the Android platform, the number of new malware samples detected by quarter has increased from about 100 in the third quarter of 2011 to over 8,000 in the third quarter of 2012 [70, 71]. Therefore, it seems unlikely that this trend will be interrupted any time soon. The trend is also confirmed by the most recently published vulnerabilities and publicly available exploits for the Android platform:

- ObjectInputStream Privilege Escalation (CVE-2014-7911), published in Nov. 2014, works up to Android 4.4.4,
- Fake ID [27], published in Aug. 2014, works up to Android 4.3,
- TowelRoot (CVE-2014-3153), published in Jun. 2014, works for Android 4.4,
- Master Key (CVE-2013-4787), published in Jul. 2013, works up to Android 4.2,
- mempodroid[2] (CVE-2012-0056), published in Jan. 2012, works for Android 4.0.1–4.0.3,
- Levitator[3] (CVE-2011-1352), published in Oct. 2011, works up to Android 2.3.5,
- zergRush[4] (CVE-2011-3874), published in Oct. 2011, works up to Android 2.3.3,
- GingerBreak[5] (CVE-2011-1823), published in Apr. 2011, works up to Android 2.3.3,
- KillingInTheNameOf[6] (CVE-2011-1149), ZimperLich,[7] RageAgainstTheCage,[8] Exploid, and others for earlier versions.

Soon after one vulnerability gets fixed, a new exploit for another vulnerability is published. If this trend continues, it is only a matter of time until exploits for the most recent versions of the Android platform become available. With mobile operating systems like Android, there is an issue that has a significant impact on the viability of exploits: There is a significant delay between fixing a vulnerability in the Android open source project and distribution of the updated Android platform to existing devices. There is even a chance that older devices do not receive updates at all.

Höbarth and Mayrhofer [47] introduce a framework for the Android platform that can use arbitrary exploits to achieve permanent privilege escalation. Based on any existing or future exploit that gains temporary *root* level privileges, their framework modifies the system in such a way that these *root* level privileges are permanently retained. Such frameworks provide an easy-to-use platform for attackers to integrate the newest exploit code into their malware applications.

Höbarth [46] uses a loophole in Google Play Store (formerly Android Market) to publish a malicious app under the same publisher information and the same app information as an existing app. This scenario can be used to trick users into mistakenly installing the malicious app instead of the genuine app. He further explains that an attacker could even reengineer an existing app downloaded from Play Store to add malicious code to a legitimately looking app.

---

[2] https://github.com/saurik/mempodroid.
[3] http://jon.oberheide.org/files/levitator.c.
[4] http://forum.xda-developers.com/showthread.php?t=1296916.
[5] http://c-skills.blogspot.com/2011/04/yummy-yummy-gingerbreak.html.
[6] http://c-skills.blogspot.com/2011/01/adb-trickery-again.html.
[7] http://c-skills.blogspot.com/2010/12/zygote-trickery-743c-27c3-release.html.
[8] http://c-skills.blogspot.com/2010/08/droid2.html.

4.6 Combining NFC with Trusted Platform Concepts

Madlmayr [68] explains that trusted computing is an essential requirement for NFC applications. Secure storage and execution of applications on a mobile device are the basis for contactless payment transactions [68].

Ekberg and Kylänpää [22] give an overview over the mobile trusted module (MTM), a trusted platform module for mobile devices. MTMs can be used to guarantee the integrity of mobile device firmware (boot a trusted boot loader, run a trusted hypervisor, run a trusted operating system, etc.). Though, Kostiainen et al. [66] note that modern smart phone operating systems are usually too large to guarantee that they are free of vulnerabilities.

Ekberg et al. [21] explain how trusted platform architectures on mobile devices can be used to create an inexpensive, open and secure alternative to secure elements. Their "on-board credentials" (ObC) provide a secure execution environment and secure storage based on secure hardware like Texas Instruments M-Shield or ARM TrustZone. Thus, they can be used to store digital keys, tickets or even payment applications. ObCs provide similar properties as conventional secure elements—for instance isolation of different credential applications and secure provisioning of credentials. At the same time, ObCs have the advantage that they do not rely on a dedicated secure element chip and that anyone can deploy new credentials to an ObC without approval of the application. However, Kostiainen [65] explains that "traditionally, only smart cards and TPMs have achieved commonly accepted security certification levels". He further notes that "lack of formal certification can be a significant problem for deployment in financial services [...]" Nevertheless, ObC platforms could be an interesting opportunity for implementing NFC applications that are less restrictive about security certification (e.g. public transport ticketing). Also, despite these concerns, Proxama [82] created a credit card application based on ARM TrustZone for secure ("card present") payment in a mobile phone web browser. Their application stores the credit card in a protected location and provides a secure user interface, both by the help of ARM TrustZone.

Recently, Dmitrienko et al. [19] proposed a security architecture for NFC-enabled mobile phones to improve the protection of the secure element against software-based attacks. Their scheme moves access control decisions for secure element access to a trusted compartment on the application processor, which leads to access control enforcement by a trusted component. Consequently, this seems to be a promising approach to mitigate the attack scenarios against the secure element discovered in this thesis.

4.7 Flaws in Existing Mobile Wallet Implementations

Zefferer [97] gives an overview of recent NFC-based payment systems and mobile wallets. While some of them are proprietary (and often closed) solutions, others are based on EMV credit card standards (e.g. MasterCard PayPass and Visa payWave).

One wallet implementation has recently gained particular interest: Google Wallet. This application is the first NFC-based mobile wallet that has become publicly available in the USA. It has quickly spread to a broad user-base. In late 2012, the app already counted over a million installations. As Google Wallet is based on MasterCard PayPass it can be used worldwide with any merchant that accepts contactless MasterCard transactions [98].

A first in-depth analysis of Google Wallet is performed by Hoog [48]: He reveals several weaknesses of the app. For instance information about the credit card holder, expiration date, some digits of the credit card number, the credit card balance and even information about transactions is stored on the device unencrypted. Information that has been explicitly deleted through the user interface of the app can even be recovered. Above all, Google Wallet uses Google Analytics to collect usage data, which results in partly unencrypted information to be sent across the Internet.

Benninger [7, 53] discovered a vulnerability in Google Wallet that allows an attacker to reveal the last four digits of all credit card numbers associated with a user's Google Wallet and Google Checkout accounts.

Rubin [87] discovered that the PIN needed to access the Google Wallet app is stored as a salted SHA-256 hash on the device. As the PIN consists of only four digits, it can easily be recovered from the hash and salt information by brute-force testing of all (10,000) combinations. Allen et al. [1] even published exploit code to demonstrate this vulnerability. Rubin [87] notes that root privileges are required to access the hash and salt information. While it is frequently argued that devices that have been intentionally "rooted" by their owners are not secure anyways, Rubin [86] explains that even if users do not intentionally root their devices, malicious apps could also exploit privilege escalation vulnerabilities to elevate their privileges to "root" access.

Fannin [26] found that the Google prepaid credit card that can be used in Google Wallet is bound to each device and not to the user's Google account. As a consequence, an attacker can bypass the wallet PIN by wiping the data of the Google Wallet app (including the PIN). The attacker can then re-initialize Google Wallet on the same device. By adding a Google prepaid card, the old (device-bound) card is added to the new wallet instance. Thus, the attacker gains access to the prepaid card including any funds that were left on it. According to Zefferer [98] this vulnerability even caused Google to temporarily disable provisioning of new prepaid cards.

Apart from attacking mobile wallet implementations, there is also vulnerability research on the underlying credit card standards. Murdoch et al. [77] show a flaw in the offline PIN verification of the EMV *Chip & PIN* protocol: An attacker who has the control over a stolen credit card can trick a point-of-sale terminal into accepting any PIN for the card by means of a man-in-the-middle attack. Bond et al. [11] discovered that many automatic teller machines (ATMs) and point-of-sale terminals use weak random number generators that are simple counters, derive their random numbers from clocks, use a reduced random number space or have bad seeding. This allows an attacker to predict the random number sequence. Hence, an attacker can pre-play the transaction authentication procedure for a series of random numbers with a genuine card. Later, the pre-computed authorization codes generated in the pre-play attack can be used to skim the card using a card clone.

4.8 Summary

The previous sections gave an overview of research activities focused on the security and privacy of NFC and its underlying RFID technologies. Figure 4.1 lists the identified research topics and positions them in the context of this thesis.

From the exemplary use-cases of NFC described in Chap. 3 two major research areas have been identified:

- security and privacy aspects of "tagging" (i.e. the interaction with NFC tags), and
- security aspects of the interaction between the application processor and the secure element in card emulation mode.

In the area of tagging, research focuses mainly on finding vulnerabilities. While some research (and already some standardization) is done to create a robust and secure tagging experience, continuous reports of similar vulnerabilities in new mobile phone

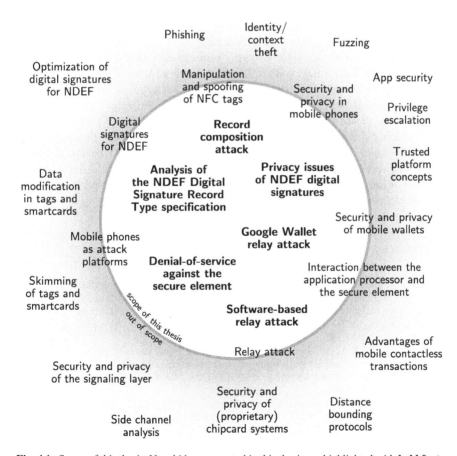

Fig. 4.1 Scope of this thesis. Novel ideas presented in this thesis are highlighted with **bold font**

platforms suggest that well-known research results are barely considered during the development of new NFC devices. This thesis follows up on these research results and analyzes the first generation of digital signature standards for the NFC data exchange format that are supposed to mitigate most vulnerabilities of tagging usecases. The results of this analysis reveal that the NDEF Digital Signature Record Type specification leaves several important parts of the signature ecosystem open to the implementer. Moreover, this record type specification contains some severe weaknesses that result in new security and privacy issues. As part of my research, a new attack scenario, the *record composition attack*, has been discovered. This scenario can be abused to manipulate integrity protected and authenticated NDEF record data.

In the area of card emulation, existing research leaves one huge gap: the mobile phone—specifically apps on its application processor—as platform for attacks against its own secure element. It is well-known that NFC-enabled mobile phones can be used to attack tags, smartcards and other NFC devices. Literature also mentions several attack scenarios that can be applied to contactless smartcards and, therefore, also to secure elements. However, with regard to the interaction between the mobile phone application processor and the secure element, literature usually assumes that the mobile phone operating system can perform trusted access control enforcement (cf. Sect. 4.4.3 and Madlmayr et al. [69]). While there is plenty of research on strategies to perform trusted access control enforcement on a mobile phone application processor (cf. Sect. 4.6), the link to securing access to the secure element is still missing. This thesis analyzes various secure element APIs (application programming interfaces) with regard to their access control mechanisms. Based on these findings two new attack scenarios against the secure element are introduced: One is a denial-of-service attack against the secure element and the other—the *software-based relay attack*—is an improved relay attack scenario that can be performed even without physical proximity to the device under attack.

References

1. Allen, J., Umadas, R., Benninger, C.: Google Wallet PIN brute forcing. Intrepidus Group Insight. http://intrepidusgroup.com/insight/2012/02/google-wallet-pin-brute-forcing/ (2012)
2. Anderson, R.: Position statement in RFID S&P panel: RFID and the middleman. In: Financial Cryptography and Data Security. LNCS, vol. 4886/2007, pp. 46–49. Springer, Berlin (2007). doi:10.1007/978-3-540-77366-5_6
3. Armando, A., Merlo, A., Migliardi, M., Verderame, L.: Would you mind forking this process? A denial of service attack on Android (and some countermeasures). In: Information Security and Privacy Research. IFIP AICT, vol. 376/2012, pp. 13–24. Springer, Heraklion (2012). doi:10.1007/978-3-642-30436-1_2
4. Attard, A.: A novel card-present payment scheme using NFC technology. Technical Report MA-2012-07, Royal Holloway University of London, Department of Mathematics. http://www.ma.rhul.ac.uk/static/techrep/2012/MA-2012-07.pdf (2012)
5. Avoine, G., Tchamkerten, A.: An efficient distance bounding RFID authentication protocol: balancing false-acceptance rate and memory requirement. In: Information Security. LNCS, vol. 5735/2009, pp. 250–261. Springer, Berlin (2009). doi:10.1007/978-3-642-04474-8_21

References

6. Barbu, G., Giraud, C., Guerin, V.: Embedded eavesdropping on Java Card. In: Information Security and Privacy Research. IFIP AICT, vol. 376/2012, pp. 37–48. Springer, Heraklion (2012). doi:10.1007/978-3-642-30436-1_4
7. Benninger, C.: Google Wallet—last four digits revealed to malware vulnerability. Intrepidus Group Insight. http://intrepidusgroup.com/insight/2012/04/google-wallet-last-four-digits-revealed-to-malware-vulnerability/ (2012)
8. Benninger, C.: Unlocking NFC deadbolts with Androids. Intrepidus Group Insight. http://intrepidusgroup.com/insight/2012/09/unlocking-nfc-deadbolts-with-androids/ (2012)
9. Benninger, C., Sobell, M.: Intro to Near Field Communication (NFC) mobile security. In: Presentation at ShmooCon 2012. Washington, DC, USA. http://youtu.be/ZWQKV0DI2jw (2012)
10. Benninger, C., Sobell, M.: NFC for free rides and rooms (on your phone). In: Presentation at EUSecWest 2012. Amsterdam, The Netherlands (2012)
11. Bond, M., Choudary, O., Murdoch, S.J., Skorobogatov, S., Anderson, R.: Chip and Skim: cloning EMV cards with the pre-play attack. Computing Research Repository (CoRR), arXiv:1209.2531 [cs.CY]. http://arxiv.org/abs/1209.2531 (2012)
12. Choudary, O., Stajano, F.: Make noise and whisper: a solution to relay attacks. In: Security Protocols XIX. LNCS, vol. 7114/2011, pp. 271–283. Springer, Berlin (2011). doi:10.1007/978-3-642-25867-1_26
13. Clark, S.: NFC Forum spec adds digital signatures to prevent tag tampering. Near Field Communications World. http://www.nfcworld.com/2010/02/11/32704/ (2010)
14. Conway, J.H.: On Numbers and Games. Academic Press, New York (1976)
15. Courtois, N.T.: The dark side of security by obscurity. Cryptology ePrint Archive, Report 2009/137. http://eprint.iacr.org/2009/137 (2009)
16. Courtois, N.T., Nohl, K., O'Neil, S.: Algebraic attacks on the Crypto-1 stream cipher in MIFARE Classic and oyster cards. Cryptology ePrint Archive, Report 2008/166. http://eprint.iacr.org/2008/166 (2008)
17. Davi, L., Dmitrienko, A., Sadeghi, A.R., Winandy, M.: Privilege escalation attacks on Android. In: Information Security. LNCS, vol. 6531/2011, pp. 346–360. Springer, Berlin (2011). doi:10.1007/978-3-642-18178-8_30
18. Desmedt, Y., Goutier, C., Bengio, S.: Special uses and abuses of the Fiat-Shamir passport protocol (extended abstract). In: Advances in Cryptology—CRYPTO '87. LNCS, vol. 293/2006, pp. 21–39. Springer, Berlin (1988). doi:10.1007/3-540-48184-2_3
19. Dmitrienko, A., Sadeghi, A.R., Tamrakar, S., Wachsmann, C.: SmartTokens—delegable access control with NFC-enabled smartphones. Cryptology ePrint Archive, Report 2012/187. http://eprint.iacr.org/2012/187 (2012)
20. Drimer, S., Murdoch, S.J.: Keep your enemies close: distance bounding against smartcard relay attacks. In: Proceedings of the 16th USENIX Security Symposium, pp. 87–102. USENIX, Boston, MA, USA (2007)
21. Ekberg, J.E., Asokan, N., Kostiainen, K., Rantala, A.: On-board credentials with open provisioning. In: Proceedings of the 4th International Symposium on Information, Computer, and Communications Security (ASIACCS '09), pp. 104–115. ACM, Sydney, Australia (2009). doi:10.1145/1533057.1533074
22. Ekberg, J.E., Kylänpää, M.: Mobile Trusted Module (MTM)—an introduction. Technical Report NRC-TR-2007-015, Nokia Research Center. http://research.nokia.com/files/tr/NRC-TR-2007-015.pdf (2007)
23. EMVCo: EMV Contactless Specifications for Payment Systems—Book C-3: Kernel 3 Specification. Version 2.1 (2011)
24. EUROSMART: Common Criteria for Information Technology Security Evaluation—Protection Profile Smart Card IC with Multi-Application Secure Platform (PP/0010). Revision 2.0 (2000)
25. EUROSMART: Smartcard IC Platform Protection Profile (BSI-PP-0002). Revision 1.0 (2001)
26. Fannin, H.: Second major security flaw found in Google Wallet...rooted or not no one is safe. The Smartphone Champ. http://thesmartphonechamp.com/second-major-security-flaw-found-in-google-wallet-rooted-or-not-no-one-is-safe-video/ (2012)

27. Forristal, J.: Android fake ID vulnerability. Talk at BlackHat US. Las Vegas, NV, USA. https://www.blackhat.com/docs/us-14/materials/us-14-Forristal-Android-FakeID-Vulnerability-Walkthrough.pdf (2014)
28. Francis, L., Hancke, G.P., Mayes, K.E., Markantonakis, K.: Potential misuse of NFC enabled mobile phones with embedded security elements as contactless attack platforms. In: Proceedings of the 1st International Workshop on RFID Security and Cryptography (RISC'09), pp. 1–8. IEEE, London, UK (2009). doi:10.1109/ICITST.2009.5402513
29. Francis, L., Hancke, G.P., Mayes, K.E., Markantonakis, K.: On the security issues of NFC enabled mobile phones. Int. J. Internet Technol. Secured Trans. **2**(3/4), 336–356 (2010). doi:10.1504/IJITST.2010.037408
30. Francis, L., Hancke, G.P., Mayes, K.E., Markantonakis, K.: Practical NFC peer-to-peer relay attack using mobile phones. In: Radio Frequency Identification: Security and Privacy Issues. LNCS, vol. 6370/2010, pp. 35–49. Springer, Berlin (2010). doi:10.1007/978-3-642-16822-2_4
31. Francis, L., Hancke, G.P., Mayes, K.E., Markantonakis, K.: Practical relay attack on contactless transactions by using NFC mobile phones. Cryptology ePrint Archive, Report 2011/618. http://eprint.iacr.org/2011/618 (2011)
32. Garcia, F.D., de Koning Gans, G., Muijrers, R., van Rossum, R., Verdult, R., Wichers Schreur, R., Jacobs, B.: Dismantling MIFARE Classic. In: Computer Security—ESORICS 2008. LNCS, vol. 5283/2008, pp. 97–114. Springer, Berlin (2008). doi:10.1007/978-3-540-88313-5_7
33. Garcia, F.D., de Koning Gans, G., Verdult, R.: Exposing iClass key diversification. In: Proceedings of the 5th USENIX Conference on Offensive Technologies (WOOT '11). USENIX, San Francisco, CA, USA (2011)
34. Garcia, F.D., de Koning Gans, G., Verdult, R., Meriac, M.: Dismantling iClass and iClass elite. In: Computer Security—ESORICS 2012. LNCS, vol. 7459/2012, pp. 697–715. Springer, Berlin (2012). doi:10.1007/978-3-642-33167-1_40
35. Garcia, F.D., van Rossum, P., Verdult, R., Wichers Schreur, R.: Wirelessly pickpocketing a Mifare Classic card. In: Proceedings of the 30th IEEE Symposium on Security and Privacy, pp. 3–15. IEEE, Oakland, CA, USA (2009). doi:10.1109/SP.2009.6
36. Gowdiak, A.: Java 2 Micro Edition (J2ME) security vulnerabilities. In: Presentation at Hack in the Box Security Conference. Kuala Lumpur, Malaysia (2004)
37. GSMA: Mobile NFC technical guidelines, version 2.0. White paper (2007)
38. Hancke, G.P.: A practical relay attack on ISO 14443 proximity cards. http://www.rfidblog.org.uk/hancke-rfidrelay.pdf (2005). Accessed Sept 2011
39. Hancke, G.P.: Practical attacks on proximity identification systems. In: Proceedings of the IEEE Symposium on Security and Privacy (S&P '06), pp. 328–333. Oakland, CA, USA (2006). doi:10.1109/SP.2006.30
40. Hancke, G.P.: Security of proximity identification systems. Technical Report UCAM-CL-TR-752, University of Cambridge, Computer Laboratory. http://www.cl.cam.ac.uk/techreports/UCAM-CL-TR-752.html (2009)
41. Hancke, G.P.: Design of a secure distance-bounding channel for RFID. J. Network Comput. Appl. **34**(3), 877–887 (2011). doi:10.1016/j.jnca.2010.04.014
42. Hancke, G.P., Kuhn, M.G.: An RFID distance bounding protocol. In: Proceedings of the First International Conference on Security and Privacy for Emerging Areas in Communications Networks (SecureComm 2005), pp. 67–73. Athens, Greece (2005). doi:10.1109/SECURECOMM.2005.56
43. Hancke, G.P., Kuhn, M.G.: Attacks on time-of-flight distance bounding channels. In: Proceedings of the First ACM Conference on Wireless Network Security (WiSec '08), pp. 194–202. ACM, Alexandria, VA, USA (2008). doi:10.1145/1352533.1352566
44. Hancke, G.P., Mayes, K.E., Markantonakis, K.: Confidence in smart token proximity: relay attacks revisited. Comput. Secur. **28**(7), 615–627 (2009). doi:10.1016/j.cose.2009.06.001
45. Haselsteiner, E., Breitfuß, K.: Security in Near Field Communication (NFC)—strengths and weaknesses. In: Workshop on RFID Security 2006 (RFIDsec 06). Graz, Austria. http://events.iaik.tugraz.at/RFIDSec06/Program/papers/002%20-%20Security%20in%20NFC.pdf (2006)

46. Höbarth, S.: Android monkeys—get it, malware it, market it. In: Presentation at Hacking Night WS 2011. Hagenberg, Austria (2012)
47. Höbarth, S., Mayrhofer, R.: A framework for on-device privilege escalation exploit execution on Android. In: 3rd International Workshop on Security and Privacy in Spontaneous Interaction and Mobile Phone Use. San Francisco, CA, USA. http://www.medien.ifi.lmu.de/iwssi2011/papers/hoebarth-spmu2011.pdf (2011)
48. Hoog, A.: Forensic security analysis of Google Wallet. viaForensics Mobile Security Blog. https://viaforensics.com/mobile-security/forensics-security-analysis-google-wallet.html (2011)
49. Hu, Y.C., Perrig, A., Johnson, D.B.: Packet leashes: a defense against wormhole attacks in wireless ad hoc networks. Technical Report TR01-384, revised Sept 2002, Rice University, Department of Computer Science. http://users.crhc.illinois.edu/yihchun/pubs/ricetr01-384.pdf (2001)
50. Hu, Y.C., Perrig, A., Johnson, D.B.: Wormhole attacks in wireless networks. IEEE J. Sel. Areas Commun. **24**(2), 370–380 (2006). doi:10.1109/JSAC.2005.861394
51. International Organization for Standardization: ISO/IEC 18092: Information technology—Telecommunications and information exchange between systems—Near Field Communication—Interface and Protocol (NFCIP-1) (2004)
52. International Organization for Standardization: ISO/IEC 13157: Information technology—Telecommunications and information exchange between systems—NFC Security (Parts 1–2) (2010)
53. Intrepidus Group: Remote enabling of verbose login can reveal last four digits of credit cards. Intrepidus Group Security Advisory. http://intrepidusgroup.com/advisories/intrepidus-advisory-20120418.txt (2012)
54. ITU-T: X.509: Information technology—Open systems interconnection—The Directory: Public-key and attribute certificate frameworks (2008)
55. Jeon, W., Kim, J., Lee, Y., Won, D.: A practical analysis of smartphone security. In: Human Interface and the Management of Information. Interacting with Information. LNCS, vol. 6771/2011, pp. 311–320. Springer, Berlin (2011). doi:10.1007/978-3-642-21793-7_35
56. Kfir, Z., Wool, A.: Picking virtual pockets using relay attacks on contactless smartcard. In: Proceedings of the First International Conference on Security and Privacy for Emerging Areas in Communications Networks (SecureComm 2005), pp. 47–58. IEEE, Athens, Greece (2005). doi:10.1109/SECURECOMM.2005.32
57. Kilås, M.: nfcsigning—Java library for signing/validation of NDEF messages. http://www.nfcsigning.org/ (2012). Accessed Sept 2012
58. Kilås, M.: Digital Signatures on NFC Tags. Master's thesis, Royal Institute of Technology (KTH), School of Information and Communication Technology, Stockholm, Sweden (2009)
59. Kim, C.H., Avoine, G.: RFID distance bounding protocol with mixed challenges to prevent relay attacks. In: Cryptology and Network Security. LNCS, vol. 5888/2009, pp. 119–133. Springer, Berlin (2009). doi:10.1007/978-3-642-10433-6_9
60. Kim, C.H., Avoine, G., Koeune, F., Standaert, F.X., Pereira, O.: The Swiss-Knife RFID distance bounding protocol. In: Information Security and Cryptology—ICISC 2008. LNCS, vol. 5461/2009, pp. 98–115. Springer, Berlin (2009). doi:10.1007/978-3-642-00730-9_7
61. Kocher, P.C., Jaffe, J., Jun, B.: Differential power analysis. In: Advances in Cryptology—CRYPTO' 99. LNCS, vol. 1666/1999, pp. 388–397. Springer, Berlin (1999). doi:10.1007/3-540-48405-1_25
62. de Koning Gans, G., Hoepman, J.H., Garcia, F.D.: A practical attack on the MIFARE Classic. In: Proceedings of the 8th IFIP WG 8.8/11.2 International Conference on Smart Card Research and Advanced Applications (CARDIS). LNCS, vol. 5189/2008, pp. 267–282. Springer, London, UK (2008). doi:10.1007/978-3-540-85893-5_20
63. Kooman, F.: Nokicert—Java X.509 certificate installation tool for Nokia phones. https://code.google.com/p/nokicert/ (2012). Accessed Sept 2012
64. Kooman, F.: Using Mobile Phones for Public Transport Payment. Master's thesis, Radboud University Nijmegen, The Netherlands (2009)

65. Kostiainen, K.: On-board credentials: an open credential platform for mobile devices. Ph.D. thesis, Aalto University, School of Science, Department of Computer Science and Engineering (2012)
66. Kostiainen, K., Reshetova, E., Ekberg, J.E., Asokan, N.: Old, new, borrowed, blue—a perspective on the evolution of mobile platform security architectures. In: Proceedings of the First ACM Conference on Data and Application Security and Privacy (CODASPY '11), pp. 13–24. ACM, Nuremberg, Germany (2011). doi:10.1145/1943513.1943517
67. Langer, J., Roland, M.: Anwendungen und Technik von Near Field Communication (NFC). Springer, Berlin (2010)
68. Madlmayr, G.: Eine mobile Service Architektur für ein sicheres NFC Ökosystem. Ph.D. thesis, Johannes Kepler Universität Linz, Institut für Computational Perception (2009)
69. Madlmayr, G., Langer, J., Kantner, C., Scharinger, J.: NFC devices: security and privacy. In: Proceedings of the Third International Conference on Availability, Reliability and Security (ARES '08), pp. 642–647. IEEE, Barcelona, Spain (2008). doi:10.1109/ARES.2008.105
70. McAfee Labs: McAfee threat report: third quarter 2011. http://www.mcafee.com/us/resources/reports/rp-quarterly-threat-q3-2011.pdf (2011)
71. McAfee Labs: McAfee threat report: third quarter 2012. http://www.mcafee.com/us/resources/reports/rp-quarterly-threat-q3-2012.pdf (2012)
72. Miller, C.: Don't Stand So Close To Me: An Analysis of the NFC Attack Surface. Briefing at BlackHat USA, Las Vegas (2012)
73. Mulliner, C.: Attacking NFC mobile phones. In: Talk at 25th Chaos Communication Congress. Berlin, Germany. http://www.mulliner.org/nfc/feed/collin_mulliner_25c3_attacking_nfc_phones.pdf (2008)
74. Mulliner, C.: Vulnerability analysis and attacks on NFC-enabled mobile phones. In: Proceedings of the International Conference on Availability, Reliability and Security (ARES '09), pp. 695–700. IEEE, Fukuoka, Japan (2009). doi:10.1109/ARES.2009.46
75. Mulliner, C.: Hacking NFC and NDEF: why I go and look at it again. Talk at NinjaCon. Vienna, Austria. http://www.mulliner.org/nfc/feed/nfc_ndef_security_ninjacon_2011.pdf (2011)
76. Mulliner, C.: Binary instrumentation on Android. In: Talk at SummerCon. New York, NY, USA. http://www.mulliner.org/android/feed/binaryinstrumentationandroid_mulliner_summercon12.pdf (2012)
77. Murdoch, S.J., Drimer, S., Anderson, R., Bond, M.: Chip and PIN is broken. In: Proceedings of the IEEE Symposium on Security and Privacy (S&P), pp. 433–446. IEEE, Oakland, CA, USA (2010). doi:10.1109/SP.2010.33
78. Nohl, K.: Cryptanalysis of Crypto-1. http://www.cs.virginia.edu/kn5f/Mifare.Cryptanalysis.htm (2008)
79. Nohl, K., Evans, D., Starbug, Plötz, H.: Reverse-engineering a cryptographic RFID tag. In: USENIX Security Symposium. USENIX, San Jose, CA, USA (2008)
80. Nohl, K., Starbug, Plötz, H.: Mifare: little security, despite obscurity. In: Talk at 24th Chaos Communication Congress (24C3). Berlin, Germany (2007)
81. Oswald, D., Paar, C.: Breaking Mifare DESFire MF3ICD40: power analysis and templates in the real world. Cryptographic Hardware and Embedded Systems—CHES 2011. LNCS, vol. 6917/2011, pp. 207–222. Springer, Berlin (2011). doi:10.1007/978-3-642-23951-9_14
82. Proxama: ARM 'Click & Pay': secure mobile wallet. In: Demo at Mobile World Congress 2012. Barcelona, Spain (2012)
83. Reid, J., Gonzalez Nieto, J.M., Tang, T., Senadji, B.: Detecting relay attacks with timing-based protocols. In: Proceedings of the 2nd ACM Symposium on Information, Computer and Communications Security, pp. 204–213. ACM, Singapore (2007). doi:10.1145/1229285.1229314
84. Rosati, T.: Elliptic curve signatures and certificates for Near Field Communications. In: Presentation at NFC Congress 2011. Hagenberg, Austria (2011)
85. Rosati, T., Zaverucha, G.: Elliptic curve certificates and signatures for NFC signature records. NFC Forum member-contributed white paper, Research In Motion, Certicom Research. http://www.nfc-forum.org/resources/white_papers/Using_ECQV_ECPVS_on_NFC_Tags.pdf (2011)

References

86. Rubin, J.: Google Wallet security: about that rooted device requirement... zveloBLOG. https://zvelo.com/blog/entry/google-wallet-security-about-that-rooted-device-requirement (2012)
87. Rubin, J.: Google Wallet security: PIN exposure vulnerability. zveloBLOG. https://zvelo.com/blog/entry/google-wallet-security-pin-exposure-vulnerability (2012)
88. Schoo, P., Paolucci, M.: Do you talk to each poster? Security and privacy for interactions with web service by means of contact free tag readings. In: Proceedings of the First International Workshop on Near Field Communication (NFC '09), pp. 81–86. IEEE, Hagenberg, Austria (2009). doi:10.1109/NFC.2009.20
89. Smart Card Alliance Contactless and Mobile Payments Council: Security of proximity mobile payments. White paper. http://www.smartcardalliance.org/resources/pdf/Security_of_Proximity_Mobile_Payments.pdf (2009)
90. Sun Microsystems Inc.: Java Card^{TM} System Protection Profile Collection. Revision 1.0b (2003)
91. Sun Microsystems Inc.: Java Card^{TM} System Protection Profile—Open Configuration. Revision 2.6 (2010)
92. Teepe, W.: Making the best of Mifare Classic (Update). http://www.cs.ru.nl/wouter/papers/2008-thebest-updated.pdf (2008)
93. Urien, P.: LLCPS. Internet Engineering Task Force, TLS Working Group. http://tools.ietf.org/id/draft-urien-tls-llcp-00.txt (2012)
94. Van Damme, G., Wouters, K.M., Karahan, H., Preneel, B.: Offline NFC payments with electronic vouchers. In: Proceedings of the 1st ACM Workshop on Networking, Systems, and Applications for Mobile Handhelds (MobiHeld '09), pp. 25–30. ACM, Barcelona, Spain (2009). doi:10.1145/1592606.1592613
95. Verdult, R., Kooman, F.: Practical attacks on NFC enabled cell phones. In: Proceedings of the Third International Workshop on Near Field Communication (NFC 2011), pp. 77–82. IEEE, Hagenberg, Austria (2011). doi:10.1109/NFC.2011.16
96. Wu, J., Qi, L., Kumar, R.S.S., Kumar, N., Tague, P.: S-SPAN: secure smart posters in Android using NFC. In: Proceedings of the IEEE International Symposium on a World of Wireless, Mobile and Multimedia Networks (WoWMoM 2012), pp. 1–3. IEEE, San Francisco, CA, USA (2012). doi:10.1109/WoWMoM.2012.6263736
97. Zefferer, T.: Konzepte und Umsetzungen NFC-basierter Zahlungssysteme. A-SIT, Studie zur Technologiebeobachtung. http://www.a-sit.at/pdfs/Technologiebeobachtung/Studie_NFC_Payments_v1.1.pdf (2012)
98. Zefferer, T.: Secure Elements am Beispiel Google Wallet. A-SIT, Studie zur Technologiebeobachtung. http://www.a-sit.at/pdfs/Technologiebeobachtung/20120803%20Studie_Google_Wallet.pdf (2012)

Chapter 5
Tagging

One of the major application scenarios of Near Field Communication (NFC) is *tagging*. The basic principle behind tagging is "it's all in a touch" [2]. This means that simply tapping an object with an NFC device immediately triggers an action. In the case of out-of-band pairing, for example, after scanning a connection handover tag with an NFC-enabled mobile phone, the phone immediately establishes a link based on the information retrieved from that tag. Similarly, a smart poster tag may convey an Internet address that is automatically opened in the mobile phone web browser, a telephone number that automatically initiates a phone call, or a ready-made SMS message that is automatically sent. While some NFC devices and some specific actions may require additional confirmation by the user, some actions may be performed automatically without any user interaction.

5.1 Security Issues

Several tagging applications already exist in the field. Most applications are based on NFC Data Exchange Format (NDEF), the standardized format for data exchange between NFC devices and for storage on NFC tags. This makes applications independent of any particular tag hardware and makes them interoperable across device platforms.

In the past, when NFC was only available on selected feature phones, applications focused mainly on the smart poster use-case and on integrating NFC into existing web-based or SMS-based ticketing and information systems. The term "smart poster" refers to posters, flyers and other advertising material equipped with NFC tags. For instance, these tags may convey an Internet address which provides further information about an advertised service, a telephone number for an advertised hotline or a ready-made SMS message for a ticket ordering service. Examples for some of the first applications based on tagging are

- the "ÖBB Handy-Ticket", a web-based train ticket in Vienna, Austria [18],
- the "Wiener Linien HANDY Fahrschein", an SMS-based e-ticket for the public transport system in Vienna, Austria [19],
- payment at Selecta vending machines in Vienna, Austria [20],
- ticketing and current traffic information for the public transport system in Gothenburg, Sweden [35], and
- traffic information and guidance for the public transport system in London, UK [36].

Today's use-cases also cover NFC-enabled business cards, connection pairing with Bluetooth devices, and the automation of arbitrary tasks on mobile phones. Examples are

- Cardolution's electronic business card solution [1, 24],
- Sony's MDR-1RBT Bluetooth headphones [33], and
- Tagstand's NFC Task Launcher for Android [34].

However, these examples are only a small excerpt of the multitude of today's tag-based NFC applications and devices.

Although, the number of available applications increases continuously, the analysis in Sect. 4.4.1 revealed that there is a significant number of security problems associated with tagging and the exchange of NDEF messages.

A serious threat is the manipulation of NFC tags. An attacker may replace (unprotected) tag content or even replace whole tags with modified tags. By, for instance, manipulating Internet addresses or telephone numbers in smart poster tags it is possible to redirect the user to a forged website for phishing of user credentials, to trigger arbitrary actions on a mobile phone in the context of the user or to trick the user into sending an SMS message to a costly premium rate service.

A practical attack is described by Mulliner [11] for payment at Selecta vending machines in Austria:

> The Selecta company started installing soda and snack vending machines that offer mobile phone payment using the paybox service. The payment works as follows: each vending machine has a unique identifier in the form of SNACK257 that is printed on the machine. A customer wishing to buy an item sends a short message containing this identifier to a phone number also printed on the machine. In the next step the machine displays that it is ready to dispense an item. After the customer selected an item the amount is charged to his paybox account. NFC-equipped vending machines feature a tag containing an SMS Smart Poster that contains the same data that is printed on the machine. The customer only needs to read the tag and send the message.
>
> A possible attack on these vending machines could have the goal of buying snacks or soda using somebody else's paybox account. The attack would work as following. The attacker produces a number of fake tags (the vending machine tags are cheap paper tags) that contain the ID of vending machine A. These are mounted on vending machines B, C, and D. The attacker only needs to wait until vending machine A shows that it is ready for selecting an item. This attack has the important advantage that it is nearly untraceable since no premium rate phone number is needed.

5.1 Security Issues

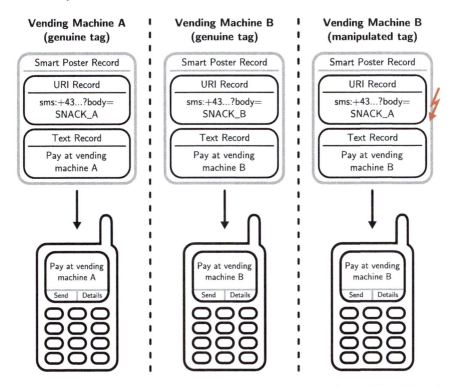

Fig. 5.1 NDEF records for a practical attack on payment at Selecta vending machines in Austria described by Mulliner [11]

The attack can even be enhanced by using the text portion of the smart poster record with a text like "Pay at vending machine B" while still using the ID of machine A in the SMS message. Figure 5.1 shows the NDEF records for this example.

An important measure against manipulation of tag content is write protection. For NFC Forum Type 1 and Type 2 tags, the tag operation specifications [14, 15] define a mechanism to permanently write-protect the data using lock bits. For NFC Forum Type 3 and Type 4 tags, the write protection mechanism is not covered by the tag operation specifications [16, 17] and is, therefore, left open to vendor specific implementations. Even some Type 2 tags (e.g. Infineon's my-d NFC) ignore the lock bits defined in the tag operation specification and use a proprietary mechanism instead. In addition, all NFC Forum tags support a soft write-protection, which is basically a flag in the tag data memory that indicates that NFC devices must not write data to the tag. However, this flag does not prevent actual write operations. For most typical applications, it is important to activate the permanent physical write protection of each tag before distribution in order to protect the tag infrastructure from malicious modifications.

Unfortunately, permanent physical write protection only prevents modification of that specific tag. An attacker could still replace the whole tag or add additional tags to the infrastructure. For instance, an attacker could use an "RFID zapper" (a device that tries to destroy a tag microchip by inducing a high voltage at the tag antenna) to disable a tag. Alternatively, a tag could be shielded with metal foil. The attacker can then stick a replacement tag with (maliciously) modified content on top of the disabled tag.

One measure to diminishing the risk of such an attack is that the NFC device verifies the authenticity and the integrity of all received NDEF records. There are two different approaches towards assuring authenticity and integrity of tag contents described in literature: Schoo and Paolucci [32] suggest that spoofing of NFC tags can be prevented by registering all genuine tags in a database back-end and by using a certified application on the NFC device that compares the data from the tags with the data stored in that back-end database. A major disadvantage of this approach is that verification requires either an online connection to the back-end database whenever a tag is scanned or an offline copy of that database. Madlmayr et al. [9] suggest digital signatures as a method to assure authenticity and integrity of NFC tags. With a combination of digitally signed NDEF messages and a trustworthy certification infrastructure, users (or their NFC equipment) have a means to distinguish genuine tags from forged tags.

5.2 Digital Signature for NDEF Messages

Digital signature consists of two cryptographic operations: calculation of a hash value and encryption. A hash function generates a fixed-length fingerprint of a variable-length message. A cryptographic hash function has three security properties [10]:

1. *Preimage resistance*: While it should be easy to calculate a hash value h from a given message M, it should be difficult to find any message M that leads to a given hash value h.
2. *Second preimage resistance*: Given a message M with a hash value h, it should be difficult to find a second message M' that leads to the same hash value h.
3. *Collision resistance*: It should be difficult to find two different messages M and M' that lead to the same hash value h.

As a result of these properties, a hash guarantees the integrity of a message.

A digital signature over a data packet is calculated in two steps: First, a hash value is calculated for the data packet. Second, the hash value is encrypted with the signer's secret key. As only the signer has knowledge of that secret key, this step assures the authenticity of the hash value. As the hash guarantees the integrity of the data packet, the encrypted hash (the "digital signature") assures the authenticity and integrity of the whole data packet.

Digital signatures based on public-key cryptography in combination with a trustworthy certification infrastructure warrant three important properties [31]:

1. *Authenticity*: The signing party can be determined unambiguously.
2. *Unforgeability*: Only the holder of a secret signing key can create an authentic signature.
3. *Non-reusability*: A signature is bound to the signed data and cannot be used for any other data. Thus, a digital signature assures the integrity of the signed data.

Therefore, properly signed NDEF data allows the receiving NFC device (and the user) to determine if an NDEF message has a certain origin and if it is free of manipulations. Based on that information a decision can be made, whether NDEF records should be allowed to trigger certain events, like opening a specific website, calling a specific telephone number or initiating a specific alternative carrier. Yet, there are several types of attacks that cannot be averted with digital signatures. Among them are the malicious modification of unlocked tags (e.g. to perform a denial-of-service attack against the tag infrastructure) and the use of valid signed tags in other than the intended places.

5.2.1 Attaching a Signature to an NDEF Message

There are several conceivable ways to attach a digital signature to an NDEF message:

- The signature could be stored in a separate memory area of the NFC tag.
- The signature could be appended (or prepended) to the NDEF message without using a separate NDEF record.
- The signature could be packed into its own record type and inserted into the NDEF message.

To stay compatible with the NDEF format and to use signatures regardless of the medium used for data exchange, a dedicated record type for digital signatures seems to be the best option. This is also the method that the NFC Forum chose for digital signature. They created the *Signature Record Type Definition* [13]. However, at the time of the research described in this part of the thesis, the Signature Record Type Definition was still under development and was, therefore, not publicly available.

5.2.2 Maintaining Backwards Compatibility

Backwards compatibility is an important requirement for digitally signing NDEF messages. There are two classes of compatibility that need to be considered:

1. *No signature support*: Devices that do not support digitally signed NDEF messages must still be able to process the NDEF message as if it had no signature. By using the approach of appending a dedicated signature record, devices that do not support signatures will simply ignore the signature as an unknown record type and will process the remainder of the message as usual.

2. *Handling of unsigned NDEF messages*: Backwards compatibility to unsigned NDEF tags is a difficult topic. On the one hand, many current applications rely on unsigned tags. Therefore, an NFC device that blocks or ignores unsigned NDEF messages would render these applications unusable. On the other hand, an NFC device should distinguish between signed and unsigned data and should use different levels of trust for each of these cases. If an NFC device would handle both cases, signed and unsigned, in exactly the same way, then the signature would be useless.

5.2.3 Signing Individual Records

A digital signature could be attached to a single record, a group of records or the whole NDEF message. Records within one NDEF message might even be issued by more than one party. Hence, signing an NDEF message as a whole with a single signature may not always be a desirable solution. The other extreme would be to sign each and every record individually. As tag memory is usually a very limited resource, this is not a reasonable solution either. Consequently, the best approach is to group the records and sign each group individually.

The Signature Record Type Definition (Signature RTD) is capable of selectively signing slices of an NDEF message by inserting begin and end markers into the message. The end marker also contains the signature for that marked slice. This method has been patented by Samsung in WO 2010/005228 A2 [29].

5.2.4 Scope of a Signature

When a digital signature is applied to an NDEF message, one could either sign each record as a whole or sign only certain fields of each record. Signing each record as a whole may lead to certain problems:

- Records may be re-formatted by the NDEF parser (e.g. the format may be changed between regular-length and short-length format, records may be split into chunks, or record chunks may be joined).
- Signed slices may be re-grouped to form new NDEF messages.

All of these cases result into changes of some bits of the signed NDEF records. Therefore, it is either necessary to prohibit such operations on signed NDEF records or to exempt certain record fields from protection by the signature.

5.2.4.1 Message Begin Flag and Message End Flag

When a signature is appended to a group of NDEF records, none of the signed records can have the message end (ME) flag set. As a result, including the ME flag results

5.2 Digital Signature for NDEF Messages   75

in an invalid signature when signing the last record of an NDEF message before actually appending the signature record.

Similarly, when signing the first record of an NDEF message (i.e. the one with the message begin (MB) flag set), that slice cannot later be moved to another position within an NDEF message as this would require the MB flag to be cleared.

In general, it is desirable to maintain the possibility of moving a group of signed NDEF records to any position within an NDEF message. Hence, including the MB flag or the ME flag into the signature is not useful.

5.2.4.2 Payload Field and Type Field

The central element to be protected by the digital signature is the record payload. Therefore, the payload field inevitably needs to be covered by the signature. As the type identification determines the interpretation of the payload field, the integrity of the type field has to be guaranteed as well.

5.2.4.3 ID Field

NDEF records may be linked to other NDEF records through their ID reference. When the ID field of a referenced record is manipulated, any such links will be broken. An attacker could use this method to bypass a record in the signed NDEF message and to redirect the link to a new record (either unsigned or signed by the attacker). Thus, the ID field needs to be part of the signed data.

5.2.4.4 Short Record Flag

The short record (SR) flag controls the size of the payload length field. When SR is set, the size is reduced from four bytes to one byte. When the signature includes neither this flag nor the payload length field, then repacking of NDEF records from one format to the other format would be possible. On the one hand, this could be used to reduce the size of an NDEF message without invalidating its signature. On the other hand, an attacker could use this feature to modify the fields that follow the payload length by moving bytes between the payload length field and the following fields. If the length fields are not part of the signature, then there is no advantage for the attacker in manipulating the size of the payload length field. Its value could be modified anyways and the three bytes could only be moved between other fields that are not covered by the signature. However, if the length fields are part of the signature, three signed bytes could be moved from the payload length into the following fields (or the other way round) without voiding the signature. Nevertheless, when this happens to change the value of the ID length or the payload length fields or when this changes the actual size of the data available for the type, ID, and payload fields, signed bytes might be missing or left over from the record. Thus, such an attack is not easily achievable. Moreover, such attacks on the payload length field could be mitigated by treating the payload length field as a 4-byte-value during signature generation and verification regardless of the value of the SR flag.

5.2.4.5 ID Length Present Flag

The ID-length present (IL) flag controls the presence of the ID length field and, consequently, of the ID field. When the signature includes neither this flag nor the length fields, then an attacker could add an ID field and use it to hide a suffix of the type field or a prefix of the payload field without invalidating the signature. Similarly, an existing ID field could be integrated into the type or the payload field. If the length fields are part of the signature, then the lengths of type, ID and payload cannot be arbitrarily chosen. Therefore, as with the SR flag, such an attack is not easily achievable in that case.

5.2.4.6 Length Fields

When the length fields are not included into the signature, then the size of the type, ID and payload fields may be changed without requiring an update of the signature. As with the IL flag, this could be exploited to move bytes between the boundaries of each of these fields. For example, parts of the ID field or even the payload field could be appended to the type field or the other way round. The signed parts of subsequent records could even be completely included into the payload field of a preceding record.

When the length fields are signed, then it becomes more difficult for an adversary to change the field sizes. An attacker could only adjust the lengths in combination with the SR flag or the IL flag. But even then the values of the length fields cannot be chosen arbitrarily.

5.2.4.7 Chunk Flag and Record Chunks

The chunk flag allows the payload of one record to be split across multiple smaller record chunks. When only the type, ID and payload fields are signed, then a signed record can be divided into chunks or merged from multiple chunks without voiding the signature. This feature is useful to join chunks to one record in order to reduce the overhead of multiple chunk headers.

Yet, permitting this feature also makes the signature prone to attacks: Even when every other field and flag, except for the chunk flag (CF) is protected by the signature, an attacker could clear a set CF to cut the remaining chunks off the record. That way parts of a chunked record payload can be chopped off. However, the remaining chunks will trigger parser errors as their type name format (TNF) field states that they continue a previous record. Only if the TNF field is also not included into the signature, an attacker could change the value of that field to "unknown" and, thereby, force the parser to ignore the trimmed chunks. In a similar fashion a subsequent record could be appended as a record chunk to the payload of its preceding record.

Using a normalized form of each record for signature generation could mitigate this scenario and, therefore, could permit rearrangement of record chunks without such vulnerabilities. Normalized form means that the signature is generated over a version of an NDEF record that is the same regardless of how the record is split across multiple chunks and regardless of any flag values that influence the structure of a record. Consequently, the full record payload and only the header of the first chunk would be used for signature calculation.

5.2.4.8 Type Name Format Field

When the TNF field is excluded from the signature, the interpretation of the type field could be changed without actually modifying the type field contents. For instance, the well-known type "urn:nfc:wkt:U" could be changed to the external type "urn:nfc:ext:U".[1]

In combination with other unprotected fields even further manipulation is possible without voiding the signature. Particularly in combination with the length fields an attacker could change the type of a record to "unknown" and integrate the unused type field into the payload (or the ID respectively).

5.2.5 Limitations of NDEF APIs

NFC devices typically provide libraries to handle NDEF records and messages. For instance, the Contactless Communication API (JSR 257, [7]) for Java ME (Java Platform, Micro Edition) includes a package for parsing NDEF messages. It may be a good approach to allow the digital signature functionality to be built on top of that API (application programming interface). This would permit a fast adoption of digital signatures for NDEF without the necessity of preceding modifications of the API.

For example, with JSR 257, it involves lengthy administrative steps to extend the API specification and it takes additional time until actual implementations are rolled out to NFC devices. Therefore, building a signature library on top of existing APIs would allow for the library to be available to application developers much faster. Unfortunately, the NDEF parser that is included into that API already puts a certain level of abstraction on the NDEF records. For example, JSR 257 automatically combines record chunks into single non-chunked records. Similarly, a typical NDEF parser API would not distinguish between regular-length and short-length records.

For a signature library on top of an existing NDEF parser API such an abstraction renders the inclusion of most header fields into the signature virtually impossible.

---

[1] Note that this external type identifier would violate the RTD specification as it does not include a domain name.

Merely header fields that only exist in the first chunk of a chunked record, like type length and ID length, can be included. Consequently, if the JSR 257 NDEF parser needs to be used, only the type, ID, payload, type length and ID length fields can be protected by the signature.

Another option would be to use a normalized form of all records (i.e. one without the concept of record chunks and without the concept of short records) for generation and verification of the signature. By signing the payload field before the normalized header fields, signature calculation can even be performed on devices with limited processing capabilities or tight memory constraints that can only handle small chunks of data at a time. As the normalized header (containing the payload length field) is processed after the payload, the length of whole record payloads need not be known before the last part of the payload is processed.

5.2.6 Recommended Practice

While some fields have to be included into the signature in order to guarantee a minimum level of integrity and authenticity, the inclusion of some fields has advantages as well as disadvantages. Yet, some other fields should never be signed. Table 5.1 gives an overview of the capabilities of JSR 257 and of the recommended scope of the signature.

A minimum of integrity and authenticity is achieved by signing the type, ID and payload fields. The MB and ME flags, however, should never be signed to allow moving blocks of signed records within an NDEF message.

Table 5.1 Record fields weighted by the benefits of not signing a field and the drawbacks through the possible attack scenarios (based on [26])

Field	Signature useful[a]	Signature possible on top of JSR 257	
		No normalization	With normalization
Message begin	--	Not considered	Not considered
Message end	--	Not considered	Not considered
Chunk flag	-	No	Not applicable
Short record flag	+	No	Not applicable
ID length present flag	+	No	Not applicable
Type name format	+	No[b]	Yes
Type length	+	Yes	Yes
Payload length	+	No	Yes
ID length	+	No[b]	Yes
Type	++	Yes	Yes
ID	++	Yes	Yes
Payload	++	Yes	Yes

[a]Weights are ++, +, - and -- (with ++ being a definitive *yes* and -- a definitive *no*)
[b]Some combinations cannot be detected

Excluding the remaining fields from the signature has several benefits: Most of these fields cannot easily be handled through the JSR 257 NDEF API. Moreover, records could be repacked to accommodate a signed message to memory requirements, even when not using a normalized record structure for signature calculation.

Nevertheless, not signing these fields opens up several vulnerabilities to attack scenarios. Some of these scenarios allow single records in the signed message to be hidden from the parser or references through the ID field to be broken intentionally without voiding a signature. These records could subsequently be replaced by new records that are either unsigned or signed by the attacker. This kind of attacks can be prevented by either signing the vulnerable fields or by putting adequate rules of authorization in place that prevent mixing signed records, unsigned records and records that are signed by multiple parties within one context.

5.3 Establishing Trust in Digitally Signed Content

When working with digitally signed data, one has to distinguish between authenticity and authorization. A digital signature on its own only provides authenticity and integrity protection. Authenticity means that the origin of a signature can be identified. Integrity protection refers to the fact that modifications of the data covered by a signature can be detected.

Based on those two properties alone, a receiver cannot determine if the signing party was also eligible to sign a specific set of data. Moreover, given only a signature, a receiver cannot determine if the signing party should be trusted at all. The problem of trust is explained by Gladman et al. [5]:

> It is particularly important to distinguish between trust in a signature and trust in the owner of a signature. Under the right conditions digital signatures can provide confidence that a person (or an entity) has signed a data item but still say nothing about the trustworthiness of the person concerned.

In other words the receiver of a signed NDEF message can take as a fact that the issuer of the signature was in possession of the secret signing key and that the signed data is unmodified. Yet, the signature alone allows no assumptions about the trustworthiness of the issuer. This is where certificates come into play. With certificates, an ultimately trusted third party certifies that the issuer of the signature can be trusted with regard to certain actions.

5.3.1 Public-Key Infrastructure

One concept to establish trust relationships to digital signatures and certificates is a public-key infrastructure (PKI). A PKI is a hierarchy of certificates that certify digital signatures based on an ultimately trusted root node. Such a hierarchy consists of certification authorities (CAs) and end-user certificates.

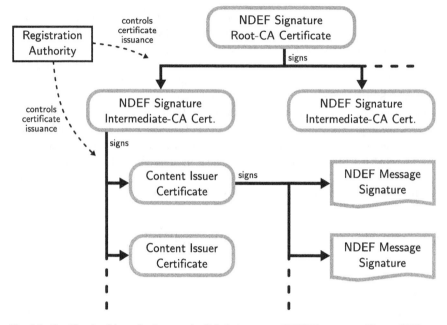

Fig. 5.2 Certification hierarchy for trust in digital signature of NDEF messages (*Source* [21])

CAs certify the identity of other CAs and end-users by digitally signing their public key and a certificate that contains information about the identity of the holder of the private key. In addition, the certificate contains information about its purpose (i.e. if the private key can be used to certify other CAs or to sign certain NDEF records). End-user certificates guarantee that the holder of a particular private key is trusted to sign certain NDEF records.

A CA has a registration authority that manages the registration of users (other CAs and end-users) and the issuance of certificates [21]. A registration authority also verifies the users' identity and their eligibility for certain certificates.

Figure 5.2 outlines a possible certification hierarchy for digital signature of NDEF messages. There must be at least one root certification authority. This root CA is the start of the chain of trust. The root CA certificate is stored in the certificate store of NFC devices and is ultimately trusted to certify intermediate CAs. Intermediate CAs that have a certificate that has been signed by an ultimately trusted root CA are trusted to issue content issuer certificates to end-users. End-users with a valid content issuer certificate that chains back to a trusted CA are trusted to sign NDEF records that match the purpose defined in their content issuer certificate.

5.3.2 Mapping Content Issuer Certificates to Content

Certificates map an issuer private key to a specific purpose of use. In the case of digital signature of NDEF messages, this could simply be the signature of any NDEF record. However, this means that anyone who is in possession of a certified private key could create any NDEF message.

Certificates have attributes to further restrict their usage. For example, secure sockets layer (SSL) certificates for web servers certify only the usage of a certain host name (or domain name). Thus, a certificate that has been issued for one server cannot be used for another server. In order to get a certificate for a certain host name, users have to prove that they are the legitimate owners of that host name.

Similarly, for NDEF records it might be desirable to bind certificates to certain information within an NDEF message. While mapping of content issuer certificates to specific content is possible with some record types, it is difficult or even impossible with other record types. In some cases binding a certificate to record content does not make sense at all [22].

Mapping is fairly easy with records that contain information in a standardized format that can be uniquely associated with an issuer (cf. [22]):

- *Uniform resource locators* (*URLs*): Signature of uniform resource identifier (URI) records with Internet addresses can be restricted to URLs that contain the host or domain name of the issuer. For instance, a certificate bound to *mroland.at* may only qualify for signing URI records that point to Internet addresses in the domain *mroland.at* and may not provide authorization for URLs in other domains.
- *SMS messages and telephone calls*: Signature of URI records with SMS messages and telephone calls can be restricted to telephone numbers that belong to the issuer. For instance, a certificate bound to the phone number *+435080427149* may only qualify for signing URI records that contain SMS messages or telephone calls for that phone number and may not provide authorization for use of other phone numbers.
- *External type records*: Signature of external record types can be restricted to type names within the issuer domain.
- *Smart poster records*: Signature of smart poster records can be restricted based on the URI record inside the smart poster.

However, in some cases a content-based mapping may not be possible or even desirable (cf. [22]):

- *Text records*: Text records do not contain a value that is suitable for mapping to a specific issuer.
- *Business cards*: While business cards (i.e. Multipurpose Internet Mail Extensions (MIME) type records of type *text/x-vcard*) contain several values that could be mapped to an issuer (e.g. the name of an organization, URLs, phone numbers or even a person's name), it is not always desirable to have such a binding. For instance, mapping the name of an organization to the certificate would work if only that organization issues business cards. However, if electronic business cards were

used at an event to allow easy exchange of contact details between participants, the business cards may be created and distributed by the event organizer and not by the organizations named within the business cards.
- *Connection Handover*: With connection handover, the hardware addresses of wireless interfaces could be used to bind certificates to specific content. There are, however, several issues with this: First, an issuer would need to get certificates that cover all of their equipment hardware addresses. Second, any issuer would get a certificate for their own wireless equipment. Thus, an attacker who wants to redirect users through their own equipment would easily get a valid certificate for handover to that equipment. Similar attacks may be possible with URIs.
- *URLs*: Even with URLs there are some situations where binding the certificate to a domain name or a host name is not feasible. For example, application developers may want to use tags to link to their application on a market place (e.g. on Google Play Store). The application developer would not be able to obtain a certificate based on that host name as it belongs to the market place and not to the developer. Using the full URL within the certificate would mitigate that problem.
- *SMS messages*: As with URLs, there are some situations where content binding based on the phone number is not feasible. An example would be a premium rate SMS phone number that is shared by multiple issuers and only distinguishes between the different services based on the SMS message text. Using the full SMS-URI within the certificate would again mitigate that problem.

In these cases, where a certificate cannot easily be mapped to the content of NDEF messages, it may make sense to let the user explicitly decide if certain issuers are allowed to perform certain actions. This is similar to code-signing certificates, where the user or the operating system decides if the apps of a certain developer are trusted to be run on a device.

As a result, content binding makes sense in some cases but is impracticable in other cases. Overall, it would make sense to have both, certificates that are mapped to certain content and certificates that are more like general code-signing certificates.

5.3.3 Partial Signatures

Binding certificates to content is not the only problem. An even sever issue arises if it is possible to sign only specific records within an NDEF message. If a signature can cover only specific records within an NDEF message, it becomes possible to mix signed and unsigned records in one message. It is even possible to create NDEF messages that contain multiple individually signed parts. The signatures may even be backed by different certificates and different issuers. The following questions emphasize the possible security issues:

- What if a smart poster contains a signed title but an unsigned URI?
- What if a smart poster contains a signed URI but an unsigned title?
- What if a smart poster title and URI are signed by different parties?

5.3 Establishing Trust in Digitally Signed Content

These questions are tightly linked to the issue of authorization. Especially the latter case may be abused to replace a smart poster URI with a malicious URI. When attackers have a valid certificate for the malicious URI they may even sign the forged part of the smart poster NDEF message. Such cases and their possible exploitation for attacks must be considered thoroughly when implementing digital signatures for the NDEF format.

Davis [4] describes three types of signing categories that exist for static messages:

- comprehensively signed content,
- (partially) unsigned content, and
- signed content groups.

Comprehensively signed content has only one signature for the whole data packet and, thus, can be trusted based on this signature. The other two categories add a potential risk to the trust relationship.

With partially unsigned NDEF messages, the receiver can only trust the signed parts of the message, while the unsigned parts have to be regarded as untrusted. With signed content groups, actions of the receiver depend on the relationship between the content groups. On the one hand, as long as the groups are unrelated, each group and its signature can be handled individually. If, on the other hand, multiple content groups share a common context, they must also share a common origin (i.e. the signatures must be issued by the same party).

An example for such a common context is the smart poster record. Its payload is an NDEF message that contains one URI record and multiple other records that describe the URI. When the smart poster record is signed as a whole (Fig. 5.3a) then the trust in the smart poster record and all its sub-records can be based on that signature and its certificate. The same applies to the case where all the sub-records of the smart poster are signed by a single party (Fig. 5.3b).

But the sub-records of the smart poster could also be divided into multiple record groups (Fig. 5.3c, d). The smart poster contains a URI record and a text record. While the URI record is signed with a signature issued by party A, the text record is signed with a signature issued by party B (for Fig. 5.3c) or has no signature at all (for Fig. 5.3d).

In the case of a text record without a signature, the text record simply cannot be trusted. However, in the case of multiple signatures issued by different parties, the evaluation of both signatures may lead to the conclusion that they are both legitimate on their own. For instance the result of this evaluation could be that the URI ("http://www.a.com/") is legitimate for A and the text ("Visit B's website") is legitimate for B. Nevertheless, in context of the smart poster record the text is a misleading description for the URI. As a consequence, the receiver has to also determine if it is safe to associate A's URI with B's text. As a general rule records in a common context, like a smart poster, should be signed by the same issuer in order to be regarded as trustworthy.

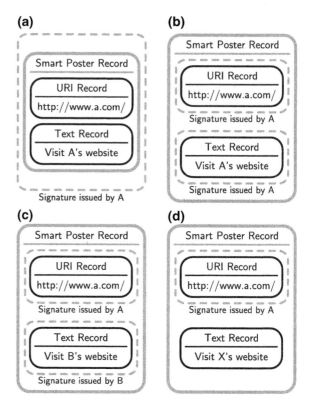

Fig. 5.3 Example: Different ways to sign a smart poster record, **a** comprehensive signature, **b** multiple signatures, same issuer, **c** multiple signatures, different issuers, **d** partially unsigned

5.3.4 Managing Content Issuer Private Keys

Plank and Kolberger [21] outline two different approaches to management of content issuer private keys: Management by the content issuer and management by the tag manufacturer.

5.3.4.1 Content Issuer

In that case, end-users manage their own secret keys. No one else has access to the keys. Thus, non-repudiation is not endangered. However, this scenario also has a negative side effect: If, for example, a content issuer wants to order a large number of tags preformatted with URLs that contain a unique serial number for each tag, then the content issuer would need to individually sign each tag URL and submit all signed data sets to the tag manufacturer. Thus, this scenario may involve huge amounts of data transferred between the content issuer and the tag issuer.

5.3.4.2 Tag Manufacturer

In that case, the tag manufacturer would manage the content issuer secret keys. As the tag manufacturer can then sign content in the name of the content issuer, this might result into a liability issue. While some tag manufacturers (e.g. those that have secure printing facilities) may have the know-how and the resources to responsibly manage these secret keys, other tag manufacturers may not have these capabilities.

The advantage of this scenario is that even for URLs with diversified content, the content issuer only needs to provide the base URL. The tag manufacturer can then diversify the URL and generate an individual signature for each tag.

5.3.4.3 Online Signature Generation

Another approach for managing the content issuer private keys is online signature generation at the CA-side. This is a centralized solution where the CA manages all the content issuer private keys. Whenever a content issuer wants to sign an NDEF message, the message (or its hash value) is transmitted to the CA signature generation service. The service then computes the signature and sends it back to the content issuer. This approach is used, for instance, with code-signing of apps for BlackBerry devices.

The main advantage of this online signature generation is that it significantly reduces the complexity at the client-side. Content issuers do not need to manage and protect their secret signing keys. When they want to delegate the production of individually personalized tags to a tag manufacturer, they could simply assign a limited number of tickets for signature generation to that manufacturer. Also, the client software needs to perform less cryptographic operations. In case a hash value is transmitted to the signature service, then the software only needs to be capable of calculating that hash value from the signed NDEF message and of attaching the received signature record to the NDEF message. In case the whole NDEF message is transmitted to the signature service, the client software neither needs to perform any cryptographic operations nor needs to be capable of parsing NDEF messages. These tasks can all be handled at the server-side.

Nonetheless, online signature generation also has some disadvantages. First, there might be privacy issues that prevent the transmission of NDEF messages to the signature service. These issues can be circumvented by transmitting hash values instead of whole NDEF messages. Even then the signature service is capable of tracking the number of issued signatures, which might not be desirable for some tag issuers.

Second, if a tag manufacturer personalizes a large number of tags with diversified content, signing that content will be time and bandwidth consuming as a separate signature has to be requested and transferred across the Internet for each tag. Also the signature service must be reliable and available at all times while a batch of NDEF messages is processed.

5.3.5 Lifespan of Certificates and Signatures

Certificates usually have a limited validity period. This validity period reflects the minimum period of time that the certified secret signing key is expected to be kept secret. While a signing key is usually expected to remain secret for a long time beyond the certificate expired, there might be some occasions where an adversary gains knowledge of that key. Typically, this would be the case if the adversary managed to gain access to the location where the secret key was stored. In the worst case scenario, weaknesses in the cryptographic algorithms or improvements in computational power would allow an adversary to compute the secret key from publicly available information (e.g. the public key, signatures). However, current cryptographic algorithms are expected to remain secure for a long time.

The validity period of certificates has a significant impact on the validity of signatures themselves. Thus, when creating a PKI for NDEF signature, it is important to analyze how long the signature on an NFC tag should remain valid. Plank and Kolberger [23] discuss four different models to cope with the lifespan of certificates and signatures: the shell model, the modified shell model, the chain model, and the modified chain model.

5.3.5.1 Shell Model

With the shell model (Fig. 5.4a), a signature is valid at a time t_{verify} if all certificates in the chain of trust are valid at that time [37]. As a consequence, a signature is considered invalid as soon as any certificate in the chain of trust expired.

For an NFC tag, this means that signed NDEF records are only valid while the content issuer certificate is valid. Therefore, the content issuer certificate needs to be valid for a reasonable time after the last NDEF message was signed with it. I.e. the certificate needs to remain valid for the whole expected lifetime of the tag.

5.3.5.2 Modified Shell Model

With the modified shell model (Fig. 5.4b), a signature is valid if all the certificates in the chain of trust were valid at the time of signature creation (t_{sign}) regardless of when the signature is verified [37]. This model has the advantage that signatures remain valid even after certificates in the chain of trust expired. However, the signature must contain a timestamp that records the time of its creation [23]. Including an accurate and unforgable timestamp in the signature would be particularly easy in combination with online signature generation as a trusted service would be responsible for determining and inserting the timestamp.

Nevertheless, this model defeats the initial purpose of certificate expiry: Signatures will be accepted as valid even after the secret key was exposed to an adversary. Even if the secret key is exposed long after the certificate expired, an adversary could easily forge the signature and back-date the timestamp to a time within the validity period.

5.3 Establishing Trust in Digitally Signed Content

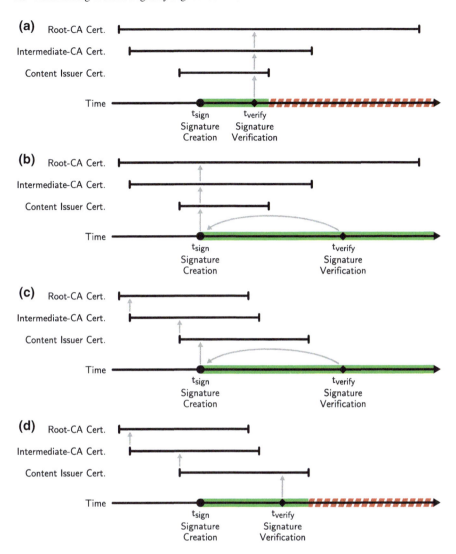

Fig. 5.4 Models for the lifetime of certificates and signatures: **a** shell model [37], **b** modified shell model [37], **c** chain model [37], **d** modified chain model [23] (based on [23])

5.3.5.3 Chain Model

With the chain model (Fig. 5.4c), a signature is valid if the certificate that certifies its signing key was valid at the time of signature creation [37]. The same applies for each certificate in the chain of trust. Similar to the modified shell model, signatures remain valid even after certificates in the chain of trust expired. Again, the signature must contain a timestamp that keeps track of its creation.

The chain model is known to be prone to attacks that allow an adversary to bypass certificate revocation lists. The vulnerability is caused by the fact that, once an adversary gains knowledge of a secret key, the timestamp included in a signature can be forged. Further information on this topic can be found in [37].

5.3.5.4 Modified Chain Model

The modified chain model (Fig. 5.4d) is similar to the chain model. However, the content issuer certificate must be valid at the time of signature verification (t_{verify}) instead of the time of signature creation (t_{sign}). This results in an equal lifetime as with the shell model. In comparison to the chain model, no timestamp needs to be included in the NDEF signature.

5.4 The NFC Forum Signature RTD

The *Signature Record Type Definition* [13] adds digital signatures to NDEF. It offers a trustworthy method for providing information about the origin of NDEF data and provides users with the possibility of verifying the authenticity and integrity of data within an NDEF message [13]. The NFC Forum released their first candidate of the signature RTD in early 2010 [3]. The candidate was adopted as a final technical specification in November 2010.

The signature RTD defines an NDEF record type as a container for digital certificates and certificate chains. It also specifies a method to attach signatures to NDEF messages.

At the time of this research, version 1.0 of the signature RTD became available. Therefore, this thesis focuses on version 1.0. Version 2.0 has been released as a candidate technical specification in 2013. That version adds stronger cryptography and tries to solve the issues discussed in this thesis.

5.4.1 Signature Record

A signature record is an NDEF record with the well-known type name "urn:nfc:wkt: Sig". The layout of a signature record payload is depicted in Fig. 5.5. The payload consists of three parts: a version byte, a signature field, and a certificate chain.

The signature field contains the signature over the signed NDEF records. The signature can either be stored directly within the signature field (denoted by a cleared URI present flag) or referenced through a URI (typically an Internet address) that points to the actual signature (denoted by a set URI present flag). The signature type field determines the algorithm that is used to calculate the signature (cf. Table 5.2, in all cases SHA-1 is used as the hash function).

5.4 The NFC Forum Signature RTD

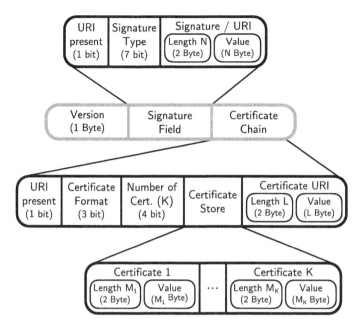

Fig. 5.5 The payload of a signature record contains the record version information, a signature part and a certificate chain (*Source* [8])

Table 5.2 Possible values for the signature type field (based on [13])

Field value	Signature algorithm
0x00	No signature
0x01	RSASSA-PSS
0x02	RSA-PKCS1-v1_5
0x03	DSA
0x04	ECDSA – P-192
Other values	Reserved for future use

The certificate chain is a list of certificates followed by an optional URI reference (if the URI present flag is set) that points to a continuation of that list. The list starts with the content issuer certificate for the signing key. Each further certificate in the list certifies the issuer of its preceding certificate. Thus, the certificates build a chain of trust. The last certificate in the list must be issued by one of the trusted root CAs. The root CA certificate itself is not part of the list. The certificate format field determines whether the certificates use the X.509 [6] certificate format (0x0) or the X9.68 format (0x1).

5.4.2 Attaching a Signature to NDEF Messages

A signature record always signs a slice of consecutive NDEF records within an NDEF message. The signature record itself is appended to that slice. Each signature record signs all preceding NDEF records starting either from the beginning of the NDEF message or from the record that follows the previous signature record. A special placeholder signature record—one with a signature type of 0x00, without an actual signature and without a certificate chain—can be used to mark the beginning of a signed slice of records while the preceding records remain unsigned. The assignment of signature records to slices of an NDEF message is shown in Fig. 5.6.

5.4.3 Signature Coverage

The digital signature contained in the signature record type does not protect all the fields of a signed record. Figure 5.7 shows how to prepare the data string for signature computation. Only the type, ID, and payload fields are covered by the signature. The remaining fields (flags, type name format, and length fields) are not part of the signed data. Consequently, those fields can be changed in signed records without voiding the signature. This allows certain operations (like changing between regular-length and short-length format, and realignment of record chunks) to be performed on signed NDEF records.

5.5 Weaknesses of the Signature RTD

In this section, the Signature Record Type Definition is further evaluated based on the analysis in Sects. 5.2 and 5.3. Several weaknesses and practical attack scenarios have been discovered.

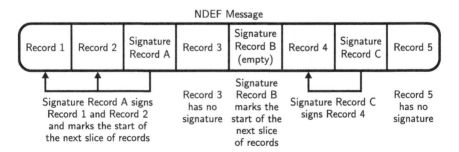

Fig. 5.6 Each signature record signs all preceding NDEF records starting either from the beginning of the NDEF message or from the record that follows the previous signature record (*Source* [8])

5.5 Weaknesses of the Signature RTD

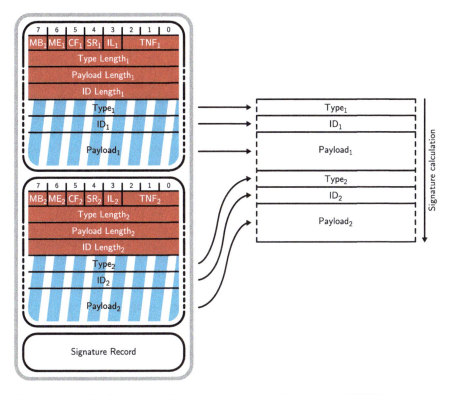

Fig. 5.7 Assembling the data string for signature calculation from the signed NDEF records

5.5.1 Establishing Trust

The Signature Record Type Definition specifies only the container format for the signature and a method to attach signatures to NDEF messages. Thus, signature records only provide integrity and authenticity. Methods for establishing trust in the legitimacy of signed data are out of the scope of the signature RTD. Implementers have to build their own PKI, and define their own policies on how to handle trust and how to establish trust relationships between content, issuers, receiving devices, and users. This makes the use of the signature RTD impractical for many real-world applications.

5.5.2 Using Remote Signatures and Certificates

A further potential weakness of the signature record type is the use of remote signatures and certificates referenced by URIs. This could open up for security vulnerabilities and privacy issues. The main problem is that the data referenced by the URIs has to be retrieved prior to verifying any signature. Therefore, the URIs within a signature record have no authenticity and integrity protection. As a result, an adversary could try to use these URIs to launch attacks.

First, if the remote URIs are accessed without notifying the user, there is a possibility of invading the user's privacy. When the URI references an Internet location, data that identifies a user (IP addresses, cookies, etc.) can be collected at a centralized service (cf. [30]) at the moment a tag is touched. This could be used to collect usage data on tags even without the need for the user to actually access the services offered by the tag. As the URIs have to be retrieved prior to the verification of the signed NDEF data, the receiver cannot make any trusted assumptions on the offered service at the time the URI resource is accessed.

An attacker could even use this approach to collect usage and user data on any existing tag infrastructure that is protected by signatures: The attacker would simply need to replace each tag with a new tag where the signature URI (or the certificate chain URI) points to a location that is controlled by the attacker. The attacker would then collect usage data whenever a signature is retrieved from the URI. In order to hide the attack from the user, the attacker would then forward the request to the original signature URI (or to the signature itself) that has been extracted from the original tag. Therefore, the user would not notice any disruption of service.

Second, the URIs are retrieved in the context of the user. As a consequence, it may be possible to use cookies and other identification data during the retrieval of the referenced URIs. An attacker could abuse this to trigger HTTP GET requests on services that are usually only available to the user. For example, the URI might trigger sending a message on an online platform (e.g. Facebook, Twitter) in the context of the user that received the NDEF message.

Furthermore, the URI could reference locations or services that are only available in the context of the receiving device. This includes resources on local network segments and on network segments that are available through virtual private network (VPN) tunnels. This also affects services that have IP address based access control and can, therefore, only be used from the user's device.

Third, the URI may be abused to trigger existing vulnerabilities of the underlying operating system. For example, this could be vulnerabilities in the URI parser or the certificate parser. In the worst case such vulnerabilities could be exploited to perform a denial-of-service or to trigger the execution of program code (e.g. through buffer overflows) on the receiving device.

5.5.3 Insufficient Signature Coverage

The signature record signs only the type, ID, and payload fields of an NDEF record. The analysis in Sect. 5.2.4 reveals that this is the worst case scenario and guarantees only a minimum of integrity and authenticity of the signed records. Yet, it allows for the use of signatures on top of certain NDEF APIs (e.g. the Contactless Communication API for Java ME).

5.5 Weaknesses of the Signature RTD

However, by changing the unsigned fields of signed NDEF records, it is possible to manipulate the semantics of signed records while still maintaining a valid signature. This leads to several potential vulnerabilities:

- Data can be moved between the three signed fields (type, ID, payload).
- Records can be hidden from processing.
- Records can be joined into a preceding record payload.
- Parts of a record payload can be extracted into separate records.

5.5.3.1 Moving Data Between Type, ID and Payload Fields

By changing the length fields, it is possible to move bytes between the type, ID and payload fields. A record with the external type "mroland.at:myapp", an empty ID and the payload "1234567890" could be changed to a record with the external type "mroland.at:my", the ID "app" and the payload "1234567890" (see Fig. 5.8). In both cases the signature is calculated over the same data string: "mroland.at:myapp1234567890". Only header fields that are not covered by the signature are changed.

5.5.3.2 Hiding NDEF Records

Record hiding can be achieved by setting the TNF field of a record to 0x5 ("unknown"). For unknown TNF, the specification [12] says:

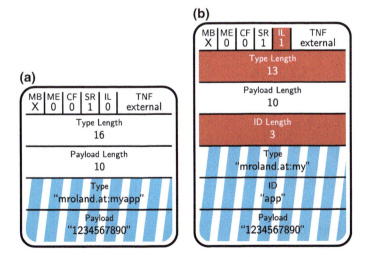

Fig. 5.8 Moving data between fields within a record: **a** original record, **b** record after moving the data (*Source* [25])

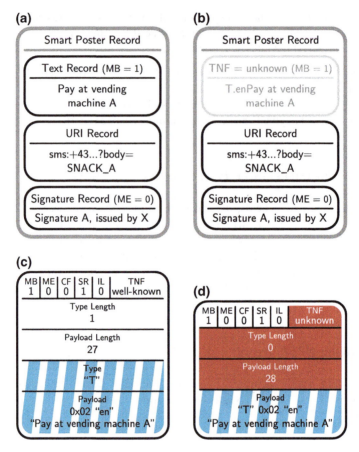

Fig. 5.9 A text record is hidden from a signed NDEF message by setting its type name format to "unknown". The type field of the text record is merged into its payload field. **a** Original smart poster, **b** smart poster with hidden text record, **c** original text record, **d** hidden text record (*Source* [25])

> Regarding implementation, it is RECOMMENDED that an NDEF parser receiving an NDEF record of this type, without further context to its use, provides a mechanism for storing but not processing the payload.

Therefore, a receiver of NDEF records should ignore such records. Consequently, an attacker has a means of selectively hiding records from signed NDEF messages without voiding the signature (see Fig. 5.9).

5.5.3.3 Joining NDEF Records

By changing the length fields of the first record and removing the header byte and length fields of subsequent records, multiple consecutive records can be joined

5.5 Weaknesses of the Signature RTD

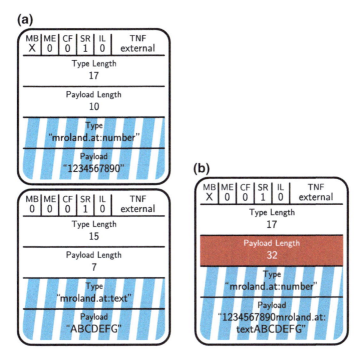

Fig. 5.10 Joining consecutive records: **a** original records, **b** joined records (*Source* [25])

into one record. For example, a record that has the external type "mroland.at:number", an empty ID and the payload "1234567890", and a record that has the external type "mroland.at:text", an empty ID and the payload "ABCDEFG" could be joined into one record with the external type "mroland.at:number", an empty ID and the payload "1234567890mroland.at:textABCDEFG". In both cases the signature is calculated over the data string "mroland.at:number1234567890mroland.at:textABCDEFG" (see Fig. 5.10).

5.5.3.4 Extracting NDEF Records

Using these methods, it is possible to extract parts of a signed record payload without voiding the signature. Unused parts can be eliminated with record hiding. Figure 5.11 shows an example: A signed smart poster record is decomposed into sub-records. One sub-record hides all unwanted parts of the original smart poster record and one record is the remaining text record. The signature is still valid after record extraction.

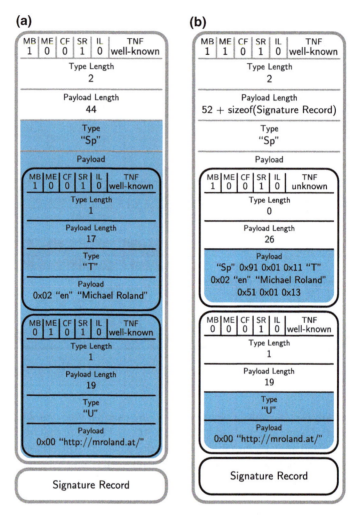

Fig. 5.11 Extracting parts of signed records: **a** original record, **b** extracted parts are embedded into a new smart poster record (*Source* [25])

5.5.4 Record Composition Attack

The combination of the above vulnerabilities leads to a new practical attack scenario: An attacker can collect multiple signed NDEF messages (e.g. from smart posters) and assemble them into a new NDEF message. The new NDEF message still contains the valid signatures of the old messages.

As explained in Sect. 5.3.3, all records that belong to a certain context need to be signed by the same party in order to establish a trust relationship between them. However, even if this rule is obeyed, there is the possibility of having multiple

5.5 Weaknesses of the Signature RTD   97

signatures from the same content issuer within one context. For example, the sub-records of a smart poster record may all have their own signature. If these signatures were issued by different parties (see Fig. 5.3c), the records can clearly not be trusted to form one smart poster record. If these signatures were issued by the same party (see Fig. 5.3b), the records may be trusted to form one smart poster record.

Therefore, if a complex record type (e.g. a smart poster record) is assembled from multiple records that were all individually signed with the same content issuer certificate, the complex record as a whole may still be regarded as validly signed. This new attack scenario is called the "Record Composition Attack" (cf. [27]).

As an example, Fig. 5.12 shows how to use the record composition attack to perform Mulliner's attack on Selecta snack vending machines (cf. Sect. 5.1) even when digital signatures are used. The attack is performed in several steps:

1. The adversary collects the smart poster records of two snack vending machines A and B. Each smart poster record contains a text record of the form "Pay at vending machine X" and a URI record with a ready-made SMS message of the form "sms:+43...?body=SNACK_X", where X is either A or B. Both smart poster records are signed by Selecta.
2. The adversary extracts the text record of the smart poster from machine A by selectively hiding the unwanted parts (i.e. the smart poster header, the URI record and the text record header) in records of "unknown" type.
3. The adversary extracts the URI record of the smart poster from machine B by selectively hiding the unwanted parts (i.e. the smart poster header, the URI record header and the text record) in records of "unknown" type.
4. The remaining records are combined into one NDEF message and used as the payload of a new smart poster record. The new smart poster record contains the text record "Pay at vending machine A", the URI record "sms:+43...?body= SNACK_B", several records of "unknown" type and a valid signature for both the text record and the URI record.

As a result, the receiver of the NDEF message will see a smart poster record containing a text record and a URI record, both with a valid signature. The attacker can now replace the NFC tag at vending machine A with a new tag containing this new smart poster message. The user will still see the expected text "Pay at vending machine A" and a certificate issued by Selecta. However, when sending the SMS message, the payment is actually processed for vending machine B as the text of the SMS message is "SNACK_B". The adversary could then wait at vending machine B until a user tries to buy a snack at machine A. As soon as a payment is authorized, the attacker can release a snack at machine B.

This scenario demonstrates that even when an NDEF message is signed by only a single party there is not necessarily a trust relationship between the signed records. Only if records are signed by the same signature record and, thus, form a single content group, they can be trusted to belong to each other.

Besides fraud, another exemplary use-case of the record composition attack is denial-of-service attacks. A denial-of-service attack can be achieved by composing

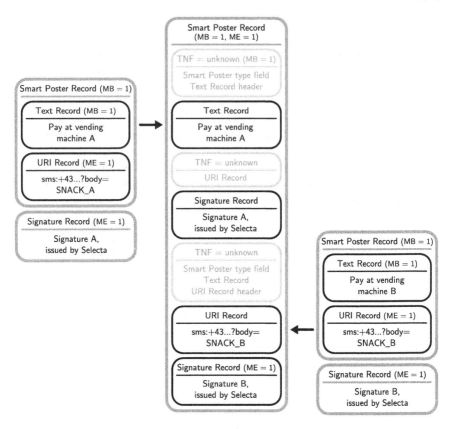

Fig. 5.12 Record composition attack: a signed(!) smart poster record conveying a new intent is assembled from parts of two signed smart poster records [27] (*Source* [25])

a message that triggers misbehavior in the receiving application. The fact that, despite the misbehavior, all records are properly signed, could lead the user into additional confusion. Mulliner [11] explains that "denial-of-service attacks can be used for destroying the trust relationship between the customer and the service provider." As the signature strongly binds the records to a certain issuer, trust in this issuer is severely endangered by such attacks.

5.6 Possible Solutions to the Discovered Weaknesses

Three weaknesses of the existing signature RTD technical specification have been discovered:

5.6 Possible Solutions to the Discovered Weaknesses

1. The signature RTD defines only a format for embedding signatures into NDEF messages. A framework for establishing trust (i.e. a public-key infrastructure) is still missing.
2. Use of URIs within signature records results in the URIs themselves not being covered by the signature. Thus, these URIs may pose a threat to users' security and privacy.
3. The signature covers only parts of the signed records and leaves certain fields unprotected. This could be abused to manipulate the semantics of signed NDEF messages without voiding their signature.

The solution to the first issue could be fairly easy as the NFC Forum only needs to define a public-key infrastructure as proposed in Sect. 5.3.

For the second issue, it could be left up to the user to decide, which URIs are trusted for certificate and signature retrieval. However, it can be assumed that the average user cannot easily distinguish between legitimate and manipulated URIs. Another possibility would be to predefine a set of allowed URIs. For instance, as with the root CA certificates, a list of trusted URIs for retrieval of intermediate CA certificates and content issuer certificates could be stored on the user's device. The content issuer certificates would then contain lists of URIs for retrieval of signatures. Nevertheless, with this solution, those service providers that have trusted URIs are still able to track users. The most restrictive solution, but also the only solution that makes this type of tracking impossible, would be to completely eliminate the usage of URI references from the signature RTD.

The third issue can only be overcome by changing the signature RTD technical specification. Attacks can only be reliably prevented if the record header fields (maybe except for MB and ME flags) are included into the signature. Of course, this would also prevent legitimate manipulations like rearrangement of record chunks and re-coding between short-length and regular-length records. A possible solution that also allows for these legitimate manipulations would be to use a normalized form of the records for calculating the signature. I.e. before calculating the signature, the NDEF message would be converted into a form that does not contain chunked or short records. Saeed and Walter [28] propose a different solution that also allows re-arrangement of record chunks by modifying the whole NDEF record format for chunked records and by including the type length and the ID length into the signed data.

In addition, to reliably prevent the composition of NDEF messages that convey a new intention from existing signed NDEF records, it is advisable to obey the following guidelines:

- The receiver of an NDEF message should only trust the relationship of records if all records are signed and if all records share one common signature record.
- The issuer of an NDEF message should sign all related records with one common signature. Unrelated records, however, should always be signed with separate signatures.

References

1. Cardolution: Electronic business card. http://www.cardolution.com/en/products/electronic-business-card/ (2012). Accessed Nov 2012
2. Chen, E.: NFC: short range, long potential. Assa Abloy FutureLab News. http://www.assaabloyfuturelab.com/FutureLab/Templates/Page2Cols____1905.aspx (2007)
3. Clark, S.: NFC Forum spec adds digital signatures to prevent tag tampering. Near Field Communications World. http://www.nfcworld.com/2010/02/11/32704/ (2010)
4. Davis, J.: Application Guidelines on Digital Signature Practices for Common Criteria Security. MSDN Magazine (2009)
5. Gladman, B., Ellison, C., Bohm, N.: Digital signatures, certificates and electronic commerce. http://jya.com/bg/digsig.pdf (1999)
6. ITU-T: X.509: Information technology—Open systems interconnection—The Directory: Public-key and attribute certificate frameworks (2008)
7. Java Community Process: JSR 257: Contactless Communication API. Version 1.1 (2009)
8. Langer, J., Roland, M.: Anwendungen und Technik von Near Field Communication (NFC). Springer, Berlin (2010)
9. Madlmayr, G., Langer, J., Kantner, C., Scharinger, J.: NFC devices: security and privacy. In: Proceedings of the Third International Conference on Availability, Reliability and Security (ARES '08), pp. 642–647. IEEE, Barcelona, Spain (2008). doi:10.1109/ARES.2008.105
10. Martin, K.M.: Everyday Cryptography: Fundamental Principles and Applications. Oxford University Press, Oxford (2012)
11. Mulliner, C.: Vulnerability analysis and attacks on NFC-enabled mobile phones. In: Proceedings of the International Conference on Availability, Reliability and Security (ARES '09), pp. 695–700. IEEE, Fukuoka, Japan (2009). doi:10.1109/ARES.2009.46
12. NFC Forum: NFC Data Exchange Format (NDEF). Technical specification, version 1.0 (2006)
13. NFC Forum: Signature Record Type Definition. Technical specification, version 1.0 (2010)
14. NFC Forum: Type 1 Tag Operation Specification. Technical specification, version 1.1 (2011)
15. NFC Forum: Type 2 Tag Operation Specification. Technical specification, version 1.1 (2011)
16. NFC Forum: Type 3 Tag Operation Specification. Technical specification, version 1.1 (2011)
17. NFC Forum: Type 4 Tag Operation Specification. Technical specification, version 2.0 (2011)
18. nfc.at: ÖBB Handy-Ticket. http://www.nfc.at/cms/front_content.php?idart=113 (2009). Accessed Nov 2009
19. nfc.at: Wiener Linien HANDY Fahrschein. http://www.nfc.at/cms/front_content.php?idart=114 (2009). Accessed Nov 2009
20. nfc.at: Zahlen am Selecta Automaten. http://www.nfc.at/cms/front_content.php?idart=37 (2009). Accessed Nov 2009
21. Plank, H., Kolberger, A.: NDEF - Signature PKI: Möglichkeiten für PKI-Infrastruktur. Projektarbeit, FH Oberösterreich, Fakultät Hagenberg, Studiengang Sichere Informationssysteme (2012)
22. Plank, H., Kolberger, A.: NDEF - Signature PKI: Verifizierungsprozesse. Projektarbeit, FH Oberösterreich, Fakultät Hagenberg, Studiengang Sichere Informationssysteme (2012)
23. Plank, H., Kolberger, A.: NDEF - Signature PKI: Zertifikatsklassen und Gültigkeitsdauer. Projektarbeit, FH Oberösterreich, Fakultät Hagenberg, Studiengang Sichere Informationssysteme (2012)

References

24. Reischl, G.: Visitenkarte 2.0 aus Österreich. futurezone.at Technology News. http://futurezone.at/b2b/2832-visitenkarte-2-0-aus-oesterreich.php (2011)
25. Roland, M.: Security and privacy issues of the signature RTD. Report to the NFC Forum security technical working group. http://www.mroland.at/fileadmin/mroland/papers/201202_SignatureRTD_Security_Issues.pdf (2012)
26. Roland, M., Langer, J.: Digital signature records for the NFC data exchange format. In: Proceedings of the Second International Workshop on Near Field Communication (NFC 2010), pp. 71–76. IEEE, Monaco (2010). doi:10.1109/NFC.2010.10
27. Roland, M., Langer, J., Scharinger, J.: Security vulnerabilities of the NDEF signature record type. In: Proceedings of the Third International Workshop on Near Field Communication (NFC 2011), pp. 65–70. IEEE, Hagenberg, Austria (2011). doi:10.1109/NFC.2011.9
28. Saeed, M.Q., Walter, C.D.: A record composition/decomposition attack on the NDEF signature record type definition. In: Proceedings of the International Conference for Internet Technology and Secured Transactions (ICITST 2011), pp. 283–287. IEEE, Abu Dhabi, UAE (2011)
29. Samsung Electronics: A NFC device and method for selectively securing records in a near field communication data exchange format message. Patent WO 2010/005228 A2 (2010)
30. Schaar, P.: Datenschutz im Internet: Die Grundlagen. Verlag C.H. Beck, München (2002)
31. Schneier, B.: Angewandte Kryptographie. Addison-Wesley, Bonn (1996)
32. Schoo, P., Paolucci, M.: Do you talk to each poster? Security and privacy for interactions with web service by means of contact free tag readings. In: Proceedings of the First International Workshop on Near Field Communication (NFC '09), pp. 81–86. IEEE, Hagenberg, Austria (2009). doi:10.1109/NFC.2009.20
33. Sony: MDR-1RBT Prestige-Kopfhörer. http://www.sony.at/product/hps-prestige-headband/mdr-1rbt/ (2012). Accessed Nov 2012
34. Tagstand: NFC Task Launcher. http://launcher.tagstand.com/ (2012). Accessed Nov 2012
35. TeliaSonera Sverige AB: TeliaSonera and Västtrafik tests new mobile technology in Gothenburg. Press release. http://www.teliasonera.com/press/pressreleases/item.page?prs.itemId=304418 (2007)
36. Transport for London: Smart posters show passengers the way. Press release. http://www.tfl.gov.uk/corporate/media/newscentre/archive/5832.aspx (2007)
37. Wölfl, T.: Formale Modellierung von Authentifizierungs- und Authorisierungsinfrastrukturen. Deutscher Universitäts-Verlag (2006)

Chapter 6
Card Emulation

While tagging is the most widely supported application scenario of Near Field Communication (NFC), card emulation is the mode that is expected to have the highest commercial impact. The reason is that, in card emulation mode, an NFC device can interact with existing contactless smartcard readers as if it were a contactless smartcard. Contactless smartcards and their corresponding reader infrastructures are already in use with several applications. For instance, more and more credit cards and credit card terminals are equipped with NFC-compatible contactless interfaces. Also, many contactless micro-payment systems and access control systems are compatible to NFC. Thus, an NFC device can operate as a payment card, as a loyalty/coupon card or as a key card for access control in these existing systems. Especially payment use-cases (e.g. credit cards) are believed to have a potential for generating high revenues.

6.1 Current Perspective on Security

The integration of smartcards as secure elements into mobile phones is seen as an advantage for both, the mobile phone and the smartcard. The mobile phone benefits from the secure storage and secure execution environment of a secure element. This could be used by apps on the mobile phone to store secret credentials or to perform secure cryptographic operations. For instance, a user could be securely authenticated to a web service using credentials stored on the secure element.

The smartcard (i.e. the secure element) benefits from the mobile phone too. It inherits all the security features of a regular smartcard (cf. Sect. 4.3) and combines them with the mobile phone user interface and network connectivity. The user interface (e.g. screen and keyboard) can be used to enhance smartcard applications with input and output capabilities. For a payment application, the user interface could provide a display to show the transaction amount and a keyboard for PIN entry and transaction confirmation. It is often assumed that the mobile phone user interface is more trustworthy than that of a payment terminal (cf. Sect. 4.4.3). In addition to simple input and output capabilities during a transaction, an app on the mobile phone could even keep track of the transaction history.

A regular (contactless) smartcard is normally disconnected from its surrounding world and only interacts with the reader infrastructure of its application. The program code and data necessary to fulfill the application requirements are pre-personalized before distribution of a card to its user. Thus, a smartcard is bound to a predefined application and can hardly be updated once it is in use.

A secure element is almost permanently connected to a global network (cellular network, Internet) through the mobile phone. This allows management of the secure element throughout its whole lifecycle. Consequently, the secure element is no longer restricted to one application. Applications can be added, configured and removed whenever the secure element has a connection to its trusted service manager (i.e. the entity that manages card applications over-the-air).

The combination of smartcards and mobile phones could even mitigate vulnerabilities. Contactless smartcards, for instance, can be accessed from a distance of several centimeters. Consequently, they are potentially prone to skimming and relay attacks. Thus, an attacker could place a reader device in proximity of a card and communicate with that card without the user's knowledge. With current contactless cards, the only viable countermeasure is to enclose the card with some protective shielding. For a secure element inside a mobile phone, there is another option: Software on the application processor could disable external card emulation whenever it is not needed. This would make access through the external interface of the secure element impossible unless the user explicitly enables it.

Literature (cf. Sect. 4.4.3) mainly covers the positive effects of combining mobile phones and smartcard technology. However, this combination can also result into new vulnerabilities. In particular, the vulnerabilities of each technology could accumulate into new and more severe vulnerabilities.

For example, while the mobile phone can restrict access to the secure element through external card emulation mode, at the same time, the mobile phone opens a new path into the secure element through internal card emulation mode. Thus, attackers might be able to access the secure element from apps or even over a network connection. Whether this potential new vulnerability poses an actual threat depends on how access control to the secure element is enforced in the mobile phone system.

6.2 APIs for Access to the Secure Element

In order to estimate how vulnerable the secure element is to unauthorized access from the application processor of a phone, it is necessary to analyze the interface between mobile phone apps and the internal mode of the secure element. Various mobile phone platforms use different application programming interfaces (APIs) and have different access control schemes for access to the internal mode of the secure element.

6.2.1 JSR 177

The Security and Trust Services API (SATSA) is standardized as JSR 177 [17] and specifies a number of Java programming interfaces for integrating secure elements into Java ME (Java Platform, Micro Edition) applications. Specifically the sub-package SATSA-APDU (`javax.microedition.apdu`) is designed for APDU (application protocol data unit) based communication (according to ISO/IEC 7816-4) with secure elements. The other packages of SATSA are SATSA-JCRMI for Java Card remote method invocation, SATSA-PKI for digital signature and credential management and SATSA-CRYPTO for basic cryptographic operations. JSR 177 is defined for Java ME, which is a Java platform specifically designed for devices with limited processing and storage capabilities (e.g. mobile phones).

In today's Java ME capable devices SATSA-APDU is mainly used for access to the universal integrated circuit card (UICC) of a mobile phone. An interface `APDUConnection` is provided for access to specific applets on the secure element. The Generic Connection Framework (GCF) is used to open connections based on the secure element slot number and the application identifier (AID) of the applet:

```
APDUConnection c = (APDUConnection)
        Connector.open("apdu:<SLOT>;target=<AID>");
```

`APDUConnection` has methods for retrieval of the answer-to-reset (`getATR`) of the card, for verification and management of PIN codes (`enterPin`, `changePin`, `disablePin`, `enablePin` and `unblockPin`) and for exchange of arbitrary APDUs with the selected applet (`exchangeAPDU`). The `exchangeAPDU` method cannot be used with commands for logical channel management and for applet selection. This limitation assures that each `APDUConnection` is bound to the initially selected applet. Regarding the PIN-related methods, JSR 177 specifies that implementations should handle PIN entry dialogs in a way that other applications can neither imitate them nor intercept entered PIN codes.

Access to the SATSA-APDU API is protected by Java ME permissions. The permissions for smartcard access are only granted to signed applications. Applications in the manufacturer domain and the operator domain are automatically granted the permission while applications in the trusted third party domain may require additional user interaction in order for the permissions to be granted.

As an addition to this basic access control scheme, the SATSA specification recommends a more sophisticated access control model in order to protect the secure element from malicious mobile phone applications. The *Recommended Security Element Access Control* [17] defines two mechanisms for fine-grained access control to secure element applications. The first mechanism extends the security domains of the Java ME device in that only applications signed with certificates that chain back to a root certificate provided by the secure element are granted access. The second mechanism is an access control scheme based on access control lists (ACLs). The secure element as a whole and each applet can have their own access control file (ACF). Each ACF contains access control entries (ACEs). The access control scheme grants access based on the APDU header information and the mobile phone application security domain (manufacturer, operator, trusted third party) or application signature (specific end-entity certificate, specific root certificate).

The SATSA specification makes some important assumptions for the access control model to be secure: Mobile phone applications are bound by all secure element access restrictions, both the mobile phone application and the applet trust the mobile device platform and only Java ME applications are considered [17].

6.2.2 Nokia Extensions to JSR 257

The Contactless Communication API is standardized as JSR 257 [18] and specifies a Java programming interface for access to contactless targets (NFC and RFID (Radio Frequency Identification) tags, contactless smartcards and visual tags). Consequently, this API provides access to NFC reader/writer mode. For their first NFC phones (specifically Nokia 6131 and Nokia 6212), Nokia developed some extensions to the Contactless Communication APIs in order to support more features of NFC. Besides support for further RFID tag types and for some limited peer-to-peer functionality, Nokia's extensions to JSR 257 also provide access to the embedded secure element of their mobile phones.

Both JSR 177 and JSR 257 provide access to smartcards. While JSR 177 is intended for access to specific applets on secure elements connected to or integrated into a mobile device, JSR 257 is intended for access to any contactless smartcard that is accessed through the NFC interface of a device. JSR 257 provides an interface `ISO14443Connection` for creating connections to contactless smartcards. The interface has a single method (`exchangeData`) to exchange arbitrary APDUs (on top of the data exchange protocol defined in ISO/IEC 14443-4) with the card. As opposed to JSR 177, a connection is not limited to one specific applet. Instead, any ISO/IEC 7816-4 APDU—including applet selection and logical channel management—can be sent to the card.

With Nokia's extensions to JSR 257 a connection can also be established to the embedded secure element of a phone. This compensates for the missing support of access to the embedded secure element through JSR 177 on their first NFC devices. An `ISO14443Connection` to the secure element is opened using the GCF. The system property "internal.se.url" contains the connection URI (Uniform Resource Identifier):

```
ISO14443Connection c = (ISO14443Connection)
        Connector.open(
                System.getProperty("internal.se.url"));
```

Opening an `ISO14443Connection` is subject to protection by Java ME permissions. On Nokia's mobile phones, however, the permission for contactless smartcard access is granted to any application by default. Thus, even applications in the untrusted third party domain (i.e. applications without a trusted signature) may freely access this API.

An additional security scheme has been introduced for secure element access through Nokia's API extensions. This scheme requires that an application is in the manufacturer, the operator or the trusted third party security domain. Therefore, only applications that are signed with trusted certificates are granted access to the secure element API.

6.2.3 BlackBerry

NFC functionality for BlackBerry devices is available since API version 7.0.0. The BlackBerry API uses the SATSA-APDU interface for secure element communication. An additional helper library (net.rim.device.api.io.nfc.se) is provided to manage multiple secure elements. The library contains the SecureElementManager singleton class for enumeration of available secure elements and for configuration of card emulation options. Each secure element is represented by a SecureElement object. This object provides methods to register for notifications about certain events (e.g. when an applet is selected through an external reader) and for retrieval of connection URIs for use with the GCF. The method getUri is used to obtain a connection URI for an APDUConnection to either a specific applet or the secure element as a whole:

```
APDUConnection cSE = (APDUConnection)
        Connector.open(se.getUri());
APDUConnection cApplet = (APDUConnection)
        Connector.open(se.getUri(AID));
```

From the available API documentation [24] it is unclear if application selection and logical channel management—which is usually not possible with an APDUConnection—are permitted if a connection has been established to the secure element as a whole. If these commands are not permitted, then only the application that is selected by default can be accessed in that case.

Access to the secure element helper library is restricted to applications that are signed with BlackBerry Java code signing keys. Code signing keys are provided to developers free of charge [22]. However, registration is required. Registrants have to provide their name, company name, country, e-mail address, website, phone number and detailed information about their project.

6.2.4 Android

While many Android-based NFC devices have an embedded secure element or support a UICC-based secure element, the Android platform API currently has no standardized interface for access to secure elements. However, several APIs exist for secure element access on today's Android devices (e.g. Samsung Nexus S, Samsung Galaxy Nexus, Samsung Galaxy S III, Sony Xperia S, etc.):

- Google's proprietary secure element API,
- SEEK-for-Android SmartCard API, and
- Open NFC secure element API.

The first two APIs are evaluated in detail in the following sections.

6.2.4.1 Google's Proprietary Secure Element API

While there is still no standardized public API for secure element access as part of the official Android Open Source Project (AOSP), Google has already introduced its Google Wallet—a secure element based container for payment, loyalty and coupon cards. Google Wallet has been available for the Nexus S in certain regions since Android 2.3.5. The wallet consists of both a mobile phone app and a component on the secure element. For the wallet to interact with the on-card component and for management of the secure element as a whole, Google secretly integrated an undocumented API called com.android.nfc_extras into their Android platform. This API can be used to access an embedded secure element and is available since Android 2.3.4. However, this interface is not included in the public software development kit (SDK) and, thus, is hidden from the average programmer.

The secure element API consists of two classes: NfcAdapterExtras and NfcExecutionEnvironment. NfcAdapterExtras is used to enable and disable external card emulation (setCardEmulationRoute) and to retrieve an instance of the NfcExecutionEnvironment class of the secure element (getEmbeddedExecutionEnvironment):

```
NfcAdapterExtras extras = NfcAdapterExtras.get(nfcAdapter);
NfcExecutionEnvironment se =
        extras.getEmbeddedExecutionEnvironment();
```

NfcExecutionEnvironment is used to establish an internal connection to the embedded secure element and to exchange APDUs with it. This class provides methods to open and close the internal connection to the secure element (open, close) and to exchange APDU sequences with the secure element (transceive):

```
se.open();
byte[] rAPDU = se.transceive(cAPDU);
se.close();
```

As opposed to other secure element APIs, the connection is established to the secure element as a whole and not bound to a single applet.

In later versions of the Android platform notifications upon external card emulation activity have been added. See Appendix A for the full interface definition.

The access control to the secure element API depends on the Android platform version:

- *Android 2.3.4*: The API can be accessed by any application that holds the permission to use NFC (`android.nfc.permission.NFC`).
- *Android 2.3.5–4.0.2*: Starting with the roll-out of Google Wallet, the permission required to access the secure element API has been changed to a special permission named `com.android.nfc.permission.NFCEE_ADMIN`. This special permission is only granted to applications which are signed with the same certificate as the NFC system service. Consequently, access to the secure element is restricted to applications that are distributed by the manufacturer/provider of the NFC system service.
- *Android 4.0.3+*: Starting with Android 4.0.3, the permission system for the secure element API has fundamentally changed. Permissions are granted in two steps. First, the Android permission system is used to verify that the app that tries to access the API has the permission to use NFC. If the app passes this check, the application certificate and package name are matched against a database of secure element access rules. Access is granted if there is a matching rule. The database is stored as an XML file (`/etc/nfcee_access.xml`). As a consequence, any application with a certificate that is listed in the XML file can gain access to the secure element. The XML file is part of the system partition and can, therefore, be updated through system updates.

6.2.4.2 SEEK-for-Android SmartCard API

SEEK, the *Secure Element Evaluation Kit* for Android, has been launched by Giesecke & Devrient as an open-source project on Google Code [7]. The project aims for creating a standard API for access to any type of secure element—the SmartCard API—that could be integrated into a future version of the Android platform. The SmartCard API started with an interface similar to the smartcard API for Java SE (Java Platform, Standard Edition). The latest version of the SmartCard API implements the Open Mobile API defined by SIMalliance (cf. Sect. 6.2.5).

While the SEEK-for-Android project submitted patches for integration into a future release of the Android platform, these patches have not been adopted as of Android 4.4. However, many devices, like the Sony Xperia S and the Samsung Galaxy S III already ship with an implementation of the Open Mobile API (typically based on the SEEK-for-Android SmartCard API).

6.2.5 Open Mobile API

The *Open Mobile API specification* [30] has been created by SIMalliance, a non-profit trade association that aims for creating secure, open and interoperable mobile services. The specification defines a platform-independent framework and an API that is not bound to any specific programming language.

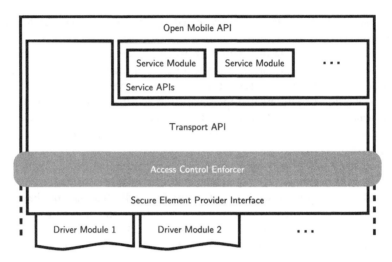

Fig. 6.1 Architectural overview of the Open Mobile API (based on [30])

The overall architecture of the Open Mobile API is shown in Fig. 6.1. The core component is the Transport API which provides APDU-based connections to secure element applets. The Transport API consists of four classes: SEService, Reader, Session and Channel. The SEService manages all secure element slots in a mobile device. Each secure element slot matches one secure element provider driver module and, thus, interfaces one secure element. Each slot is represented by an instance of the Reader class. The Reader class has methods to check the availability of a secure element and to establish a session to the secure element. Once a session is established it is represented by a Session object. The Session class provides methods to obtain the answer-to-reset (ATR) of the secure element and to open APDU-based communication channels to applets on the secure element. Each communication channel is represented by a Channel object. The Channel class has a transmit method to exchange APDUs.

The Service API consists of multiple service modules. Each module provides an application-specific abstraction of the transport layer. Thus, instead of low-level communication through APDUs, high-level methods can be defined for specific applications. For example, an authentication service API could provide methods for PIN code management and verification. Similarly, a file management service API could provide methods to create, write, and read files in a smartcard file system.

An access control enforcer between the Transport API and the secure element providers ensures that access restrictions to secure elements are obeyed. The security mechanism for access control enforcement is defined by GlobalPlatform's Secure Element Access Control specification.

6.2.6 Secure Element Access Control

GlobalPlatform's *Secure Element Access Control* [9] specification defines a sophisticated security scheme for secure element APIs to prevent secure element access from unauthorized applications. The scheme is similar to the *Recommended Security Element Access Control* of JSR 177.

The architecture of the access control scheme is depicted in Fig. 6.2. The heart of the access control scheme is the access control enforcer. The enforcer resides within the secure element API and acts as a gatekeeper between mobile phone apps and the secure element. Access control decisions are based on access control rules. Each rule defines access rights for a specific secure element applet (or for all other applets) and a specific app, a group of apps or all other apps on the mobile phone based on their certificates. Access rights can grant and deny access to all APDUs, to specific APDUs and to event notifications.

The access control enforcer reads the access control rules from a database on the secure element. Different methods for access to the database exist. The database can be a simple file, the access rule file (ARF), which is accessible through file access APDUs. The preferred way, however, is an access rule applet (ARA). As the access rule databases may be distributed across multiple security domains so that they can be managed over-the-air by multiple entities, the ARA aggregates all these databases and provides a standardized interface to the access control enforcer.

When a mobile phone app tries to access an applet on the secure element, the access control enforcer retrieves the app certificate from the mobile phone operating system (application manager) and looks up the access rules for that certificate (or its

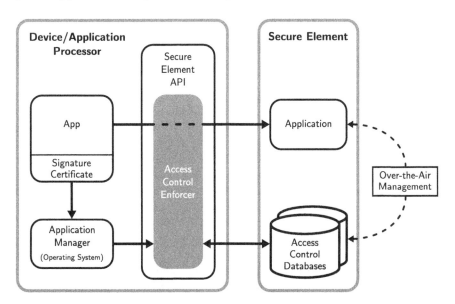

Fig. 6.2 Secure element access control architecture (based on [9])

certificate chain) and the applet AID. Based on these rules, access control is enforced for each transmitted APDU. Therefore, the access control enforcer needs to trust the operating system to provide the correct certificate that identifies the app.

6.2.7 Comparison of Access Control Schemes

The examined APIs have diverging access control mechanisms. Table 6.1 shows a comparison of these mechanisms. All schemes require an application to be signed with a valid code-signing certificate. On Android 2.3.4 this is the only requirement to request the permission for secure element access. The other schemes additionally require the code-signing certificate to be trusted by the operating system or the secure element. On Android 2.3.5–4.0.2 this trust is established by comparing the certificates of the application and the NFC service. Only if those certificates match, the application is granted access. Thus, only the system manufacturer can distribute apps that are granted this permission. With Nokia's extensions to JSR 257 any valid code-signing certificate that chains back to one of the code-signing root certificates stored on the device can be used. For the BlackBerry API, special code-signing keys, provided by the device manufacturer on request, are necessary. JSR 177 and GlobalPlatform's Secure Element Access Control have the most sophisticated access control mechanisms. Both allow the definition of access control policies that consider specific applets, the mobile phone app certificate, and specific APDUs.

Yet, with all schemes access control is enforced by the operating system on the application processor. Thus, the secure element has to assume that the underlying operating system and the mobile phone hardware can be trusted. So, in all cases, the secure element (secure component) blindly trusts the access control decisions of the operating system (i.e. the insecure component). Therefore, once an application passes or bypasses the security checks performed by the operating system, it can exchange (arbitrary) APDUs with the secure element.

Particularly, the Secure Element Access Control specification assumes that the mobile phone operating system "can be trusted about the validity of the certificates and the corresponding signatures" it provides to identify apps that trigger the access control enforcer [9]. At the same time that specification states that "restricting the use of [the secure element API] is necessary since mobile operating systems do not efficiently prevent unauthorized parties from abusing the API and potentially causing damage to the Secure Element itself." Recent research results on malicious software and attacks against mobile phone platforms (cf. Sect. 4.5) also strongly suggest that many operating systems on current mobile phones are potentially vulnerable to attacks that allow escaping the restrictions imposed by the operating system. For instance, the signature-based access restriction of the secure element API on Android 2.3.5–4.0.2 can be circumvented by abusing a vulnerability that exists throughout all of these platform versions (Master Key, CVE-2013-4787). Similarly, a vulnerability (Fake ID [3]) allows bypassing of the access control scheme of the secure element API on Android 4.0.3–4.3. As a consequence, access control enforcement within the

6.2 APIs for Access to the Secure Element

Table 6.1 Comparison of secure element access control schemes (based on [27])

	JSR 177 basic access control	JSR 177 recommended access control	JSR 257 Nokia extensions	BlackBerry	Android 2.3.4	Android 2.3.5–4.0.2	Android 4.0.3+	Globalplatform secure element access control
Valid certificate required	●	●	●	●	●	●	●	●[a]
Trusted certificate required	●	●	●	●	–	●	●	●
Manufacturer signature required	–	–	–	–	–	●	–	–
Connection on a per-applet basis	●	●	–	●[b]	–	–	–	●[c]
APDU-based access rules	–	●	–	–	–	–	–	●
Access control policy resides on the secure element	–	●	–	–	–	–	–	●
Access control enforcement by the operating system on the application processor	●	●	●	●	●	●	●	●

[a]Depends on the access rules and the operating system
[b]Assumption only
[c]In combination with Open Mobile API

mobile phone operating system may not sufficiently protect the secure element as an application might be able to bypass these security measures.

6.2.8 Impact of Rooting and Jail Breaking

For many mobile phone platforms there exist methods that users can use to intentionally circumvent security measures. Popular techniques used on many smart phones are "jail breaking" and "rooting". Jail breaking refers to escaping the restrictions imposed by the operating system, so that an application can access resources it usually could not access. Rooting refers to a slightly different scenario where the user or an application gains full access to the whole system. Both methods are often used intentionally by device owners/legitimate users to circumvent digital rights management or to gain "improved" control over their device.

However, these elevated privileges are exactly what is necessary to circumvent secure element access control enforcement. Thus, jail breaking and rooting imposes a significant security risk to the secure element. Even worse, not only the legitimate user may gain access to—otherwise restricted—resources but also an attacker could get these same possibilities. Thus, a jail broken or rooted phone is significantly more vulnerable to attacks.

On the one hand, rooting can be done by using vendor-supplied methods. Such methods typically exist for development phones (e.g. for the Google Nexus series of Android smart phones). They are usually implemented in a safe way that protects the user from malicious activities. For example, rooting a Google Nexus phone according to the official instructions will wipe all data on the phone. Thus, this method of rooting cannot be used to gain access to sensitive user data that resides on the device.

On the other hand, jail breaking and rooting can be done by exploiting vulnerabilities in software or hardware (cf. Sect. 4.5). Unfortunately, these exploits are not only viable for intentional jail breaking and rooting by the device owners/legitimate users. The same exploits can be integrated in virtually any application. That way a malicious application could elevate its permissions even without the (legitimate) user's knowledge. Considering the current continuous trend in privilege escalation exploits for various mobile device platforms (cf. Sect. 4.5), it can be assumed that an arbitrary application can use exploits to bypass restrictions and security checks performed by the operating system on most platforms that are currently in the field. Thus, an attacker may easily bypass secure element access restrictions on those devices.

6.3 New Attack Scenarios

Under the assumption that arbitrary applications may gain full control of the mobile phone operating system, the secure element is not sufficiently protected against attacks from such applications. Therefore, it can be assumed that applications can get unrestricted access to the internal mode of the secure element. As a result, attacks that were mitigated by the fact that it is difficult to secretly interface the secure element through its external RF (radio frequency) interface could be performed by an app at any time while the mobile phone is turned on. These apps can even connect the secure element to the cellular network and the Internet.

Two possible attack scenarios have been analyzed in this thesis: denial-of-service and relay of communication. Based on this analysis, two new practical attacks have been found:

- a denial-of-service (DoS) attack through the GlobalPlatform card management interface and
- a software-based relay attack on secure element applications.

The attack scenarios can be applied to the existing NFC-enabled mobile phones Nokia 6131, Nokia 6212 (both running the latest firmware), Samsung Nexus S and possibly to other devices too.

6.3.1 Denial-of-Service (DoS)

The first attack scenario is a denial-of-service attack that can be used to render a secure element temporarily unusable or to permanently prevent future card content management.

6.3.1.1 GlobalPlatform Card Management

Secure elements must be manageable while they are in the field. Certain types of secure elements do not need the application processor to establish a management channel. For instance, on a UICC, management messages could be processed immediately as they arrive over the mobile phone network and need not be passed through the application processor. However, for most secure elements the management interface is exposed to the application processor so that over-the-air management is possible using the wireless communication interfaces of the mobile phone.

An interface for card content management that is used for most of today's secure elements is standardized in the GlobalPlatform Card Specification [8]. A GlobalPlatform compliant secure element contains a *Card Manager* which consists of the *GlobalPlatform Environment* (OPEN), the *Issuer Security Domain* (ISD) and the cardholder verification method services. The card manager is the central component used to manage card content (applications and data), supplementary security domains and the whole card lifecycle.

In order to access the card manager, a service manager has to authenticate and establish a secure channel to the security domain. A sequence of three APDU commands is used for mutual authentication and to establish a shared secret for a secure channel:

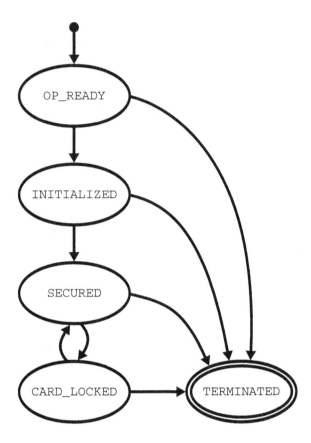

Fig. 6.3 Card lifecycle according to GlobalPlatform (based on [8])

1. SELECT security domain by AID (to select the card manager),
2. INITIALIZE UPDATE (to start the authentication procedure and to authenticate the card/secure element), and
3. EXTERNAL AUTHENTICATE (to authenticate the service manager/host).

Once a secure channel to the issuer security domain is established, the service manager can load, install, personalize and uninstall applications and supplementary security domains, update application data and manage the card lifecycle.

GlobalPlatform defines five states for the lifecycle (Fig. 6.3) of a smartcard [8]:

- OP_READY,
- INITIALIZED,
- SECURED,
- CARD_LOCKED, and
- TERMINATED.

OP_READY is the initial state after card production. During initialization with initial keys for card management, the lifecycle state irreversibly traverses from OP_READY via INITIALIZED to SECURED. In the state SECURED, the card is ready for issuance. When the card is in the state CARD_LOCKED, applications on the card can be used, but card content can no longer be managed (i.e. no applications can be added or removed). This state is reversible to SECURED. TERMINATED is similar to CARD_LOCKED, but transitions to this state are irreversible. Thus, once a card reached the TERMINATED state, management of card lifecycle and card content are no longer possible. As this state is permanent it is intended for cases where a severe security threat was detected or where a card has expired [8].

6.3.1.2 Irreversible Denial-of-Service

Many contactless smartcards and secure elements have a special security scheme to protect the card manager from unauthorized access: After a certain number of successive authentication failures, the card is put into TERMINATED state. This mechanism will prevent brute-force attacks on the authentication keys and, thus, makes the card more secure.

However, an attacker who can send APDUs to the secure element can abuse this security mechanism to permanently block a secure element from further card management. Thus, an attacker can mount a denial-of-service attack by repeatedly issuing the three commands needed for an authentication attempt—SELECT (issuer security domain), INITIALIZE UPDATE, and EXTERNAL AUTHENTICATE—until the card transitions into TERMINATED state.

Many contactless smartcards and the secure elements embedded into Nokia's first NFC phones (i.e. Nokia 6131 NFC and Nokia 6212) transition into TERMINATED state after only ten successive authentication failures (cf. [4]). Consequently, at most 30 APDU commands are needed to render an NFC device unusable for new card emulation applications. This makes it relatively easy to create an app for performing that type of denial-of-service. Malicious code for permanently locking the secure element could even be injected into any (harmless looking) application (e.g. a game).

As a result, this type of denial-of-service attack on the card manager may lead to a significant decrease in reputation and user satisfaction. Moreover, it might result in costly product recalls.

6.3.1.3 Temporary Denial-of-Service

Newer secure elements have improved that security scheme. For example, the secure element embedded into the Samsung Nexus S allows 47 authentication attempts until it transitions into TERMINATED state. However, this is not the only improvement. After five successive authentication failures, the secure element slows down the communication speed by adding a penalty of 30 s before generating a response. As a consequence, it is not easy for a malicious application to trigger a permanent denial-of-service.

However, the delayed response by the secure element could be abused to perform a temporary denial-of-service attack: Many secure elements can only handle one of the two communication modes (internal mode and external mode) at a time. Thus, if the secure element is blocking on a response to an authentication attempt, it cannot be used in external mode. Similarly, while one command-response sequence is processed, it is unlikely that the secure element will accept another command. Thus, it is also not accessible from other apps in internal mode. An attacker could use this to render the secure element temporarily unusable for the duration of the authentication attempt.

6.3.2 Software-Based Relay Attack

Hancke [11] concludes about the relay attack on contactless smartcards:

> If a contactless card could be read while in a pocket, purse or wallet, a thief might be able to engage in the act of digital pickpocketing while standing next to or merely walking past his victim.

This type of attack is often mitigated by the short reading distance of ISO/IEC 14443-based cards that makes it difficult to access a smartcard or a secure element-enabled mobile phone that is inside the victim's pocket. Moreover, mobile phones would usually turn the contactless interface off when they are not in use.

However, things are different if the internal interface of the secure element can be used for the attack. Instead of being physically close-by to the victim's phone, the attacker simply needs to install an application onto the victim's mobile phone. As with the denial-of-service attack, relevant exploit code could be packed into virtually any application.

6.3.2.1 Relay Attack on Contactless Smartcards

The relay attack is a well-known issue with contactless payment cards and secure element-enabled devices. It has been first evaluated in the context of contactless smartcards by Hancke [11], Kfir and Wool [19].

A relay attack can be seen as a simple range extension of the contactless communication channel (see Fig. 6.4). Therefore, an attack requires three components:

1. a reader device (also called *mole* [11] or *leech* [19]) in close proximity to the card under attack,
2. a card emulator device (also called *proxy* [11] or *ghost* [19]) that is used to communicate with the actual reader, and
3. a fast communication channel between these two devices.

The attack is performed by bringing the mole in proximity to the card under attack. At the same time, the card emulator is brought into proximity of a reader device (e.g. point-of-sale (POS) terminal, access control reader). Every command that the card emulator receives from the actual reader is forwarded to the mole. The mole,

6.3 New Attack Scenarios

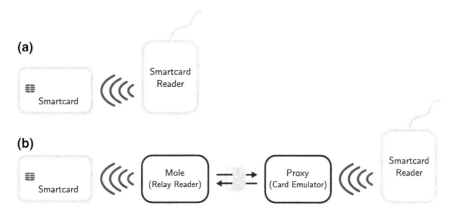

Fig. 6.4 Interaction between a smartcard and a reader: **a** without relay, **b** with relay (*Source* [25, 29])

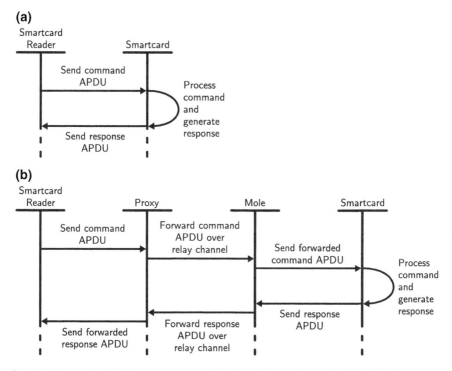

Fig. 6.5 Communication between a smartcard and a reader: **a** without relay, **b** with relay

in turn, forwards the command to the card under attack. The response of the card is then received by the mole and sent all the way back through the card emulator to the actual reader. Figure 6.5 shows the command flow without relay (a) and with relay (b).

This type of attack cannot be prevented by application-level cryptography [11, 12]. The problem is that the relay attack is a simple range extension of the contactless interface, so neither the mole nor the card emulator need to "understand" the actual communication. They simply proxy any bits of data they receive. As a consequence, encrypted and integrity protected data packets can simply be forwarded between the reader and the card. Both, the reader and the card will not notice any difference. Even current EMV credit card payment protocols using Mag-Stripe mode as well as EMV mode (*Chip & PIN*) can be relayed.

While initial approaches to relay attacks [11, 19] focused on forwarding physical layer protocols (bit transfer level), recent approaches [5, 6, 27, 28] skip the lower layers and directly transfer the APDUs of the application layer protocols. This relaxes timing requirements and greatly improves achievable relay distances.

Since their initial proposal, the practicability of relay attacks has greatly improved due to the availability of NFC-enabled mobile phones. Francis et al. [5] showed that it is possible to relay NFC signals over Bluetooth using two mobile phones. They [6] further revealed that, with the introduction of software card emulation in some smart phones, it is even possible to relay contactless credit card transactions and electronic passport transactions between two phones. Thus, NFC-enabled mobile phones can be used as both mole (relay reader) and proxy (card emulator).

6.3.2.2 The Next Generation: Software-Based Relay Attack

The threat potential of relay attacks was mitigated by the fact that all relay scenarios required close physical proximity to the device-under-attack. However, this thesis follows a different approach. Instead of accessing the secure element of a device through the contactless interface (external mode), the internal mode is used. Thus, the secure element is accessed from an app running on the application processor of the device. While the original relay attack required mole hardware in physical proximity of the device-under-attack, pure software (malware) on an attacked device application processor replaces the physical mole.

The complete relay system has been first suggested in [27] and has been verified for the Samsung Nexus S in [28]. Figure 6.6 gives an overview of the software-based relay system. It consists of four parts:

1. a mobile phone (under the control of its owner/legitimate user),
2. a relay software (under the control of the attacker),
3. a card emulator (under the control of the attacker), and
4. a reader device (under the control of its owner, e.g. at a point-of-sale terminal or at an access control gate).

The relay software is installed on the victim's mobile phone. This application is assumed to have the privileges necessary for access to the secure element and for communicating over a network. These privileges can be either explicitly granted to the application or acquired by means of a privilege escalation attack. The relay application waits for APDU commands on a network socket and forwards these

6.3 New Attack Scenarios

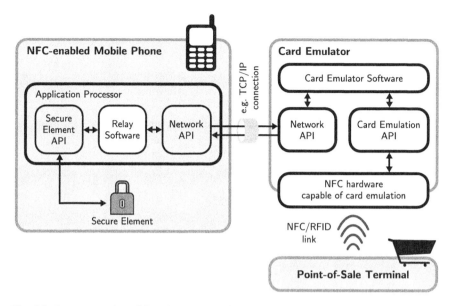

Fig. 6.6 System overview of the relay scenario (*Source* [25, 28])

APDUs to the secure element. The responses received from the secure element are then sent back through the network socket.

The card emulator is a device that is capable of emulating a contactless smartcard in software. The emulator has RFID/NFC hardware that acts as a contactless smartcard when put in the RF field of a smartcard reader. The emulator software forwards the APDU commands (and responses) between a network socket and the RFID/NFC emulator hardware.

The flow of relayed smartcard commands (APDUs) between the "real" smartcard reader and the secure element is shown in Fig. 6.7. The command APDUs (C-APDUs) received from the reader device (here: point-of-sale terminal) are routed through the card emulator and over a wireless network (e.g. cellular, Wi-Fi, Bluetooth) to the victim's device. There, the relay app forwards the C-APDUs to the secure element. The corresponding responses (R-APDUs) generated by the secure element are routed all the way back—through the relay app, the wireless network and the card emulator—to the "real" reader device. As a result, the point-of-sale terminal would believe that it talks to the secure element directly.

In comparison to existing relay scenarios by Hancke [11], Kfir and Wool [19], where bits are relayed at the data link layer, our attack scenario relays command and response packets (APDUs) at the application layer. Due to this high protocol level, there are practically no timing constraints on the delay through the relay channel. Therefore, Bluetooth, Wi-Fi, or even the mobile phone network could be used as a relay channel.

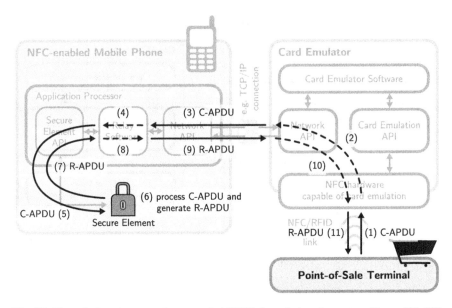

Fig. 6.7 Flow of relayed smartcard commands (APDU) through the relay system (*Source* [25, 29])

6.4 Viability of the Software-Based Relay Attack

The software-based relay attack seems to be a promising scenario from an attacker's point of view. Nevertheless, it is only a theoretical concept. Further evaluation is necessary to verify the usability of this concept in a real-world system.

6.4.1 Constraints of the Protocol Layers

There are several protocol layers and several specifications and standards involved in smartcard communication. For instance, a MasterCard PayPass credit card transaction uses protocols defined in the MasterCard PayPass specifications, in the EMV specifications for payment systems, in the ISO/IEC 7816 series of standards and in the ISO/IEC 14443 series of standards.

6.4.1.1 ISO/IEC 14443

ISO/IEC 14443 specifies multiple delays and timeouts. A detailed summary of these timings has been published by Issovits and Hutter [16].

First, there is the *frame delay time* (FDT) which is defined in ISO/IEC 14443-3 [15] for the Type A protocol. The FDT is the time between commands sent by the reader

6.4 Viability of the Software-Based Relay Attack

and responses sent by the card. For commands that are used during anti-collision, the FDT defines a strict timing between a request and the following response. This is necessary to assure synchronicity between all cards during the bit-frame anti-collision procedure in order for the reader to detect collisions. For all other commands and for the delay between a response from the card and the next request by the reader, the FDT specifies only a minimum delay. Similarly, ISO/IEC 14443-2 [14] and ISO/IEC 14443-3 [15] specify timing constraints for the Type B protocol. These are minimum and maximum delays between a request from the reader and the response from the card (TR0, TR1) and the minimum delay between a response from the card and the next request from the reader (TR2).

Second, there is the *frame waiting time* (FWT) defined in ISO/IEC 14443-4 [13] for the half-duplex block transmission protocol. The FWT specifies the maximum timeout between a block sent by the reader and the corresponding response block returned by the card. This timeout is defined by the card and can range between about 302 μs and 4,949 ms. The timeout can be extended indefinitely (for each block sequence) by the card using *frame waiting time extension* (WTX) blocks.

Hancke et al. [12] conclude that these timing constraints are too loose to provide adequate protection against relay attacks. Particularly, the software-based relay attack only has APDU-based access to the secure element. As a consequence, the lowest protocol layer that is transferred across the relay channel is the APDU layer. As the APDU protocol sits on top of the ISO/IEC 14443-4 half-duplex block transmission protocol, the ISO/IEC 14443 protocol and timing must be handled by the card emulator anyways. Therefore, it is not affected by the relay channel. Consequently, the card emulator could use the frame waiting time extension mechanism to prevent timeouts until a response is supplied on the relay channel (cf. Issovits and Hutter [16]). Even if the card emulator would not use frame waiting time extension, relaying an APDU may take almost 5 s without violating the FWT timeout.

Also an application note by MasterCard on transaction optimization for PayPass—M/Chip terminals [20] states that "it is not unusual for many [frame waiting time extensions] to be required to complete a transaction [...]". Thus, in particular credit card terminals support and even expect long waiting time extensions. Some systems, like electronic passports (machine-readable travel documents), use the waiting time extension mechanism to artificially insert long delays into the communication after failed authentication attempts to diminish the efficiency of brute-force attacks.

6.4.1.2 EMV Payment Systems

The EMV specification for contactless payment systems [2] specifies a limit of 500 ms for a contactless payment transaction as a whole. In the sense of that specification a *transaction* is the total time that a card needs to be in proximity to the reader. Thus, this includes processing times on both the terminal and the card, anti-collision, selection, and multiple APDU command-response sequences.

However, a payment terminal is not required to interrupt a transaction if it takes longer than this limit. The limit is merely meant as a benchmark target to maintain user experience. For example, the PayPass terminals used in recent roll-outs in Austria (Hypercom Artema Hybrid combined with a ViVOpay 5000 contactless terminal) do not enforce any such timings. In fact, tests revealed that transactions could take more than 20 s without being interrupted. Francis et al. [6] measured that the EMV payment terminal they used for relay attacks interrupts a transaction after 35 s.

Also, cloud-based secure element solutions (cf. [26, 34, 35]) like those provided by YES-wallet [39] require transactions to be processed over the Internet ("online"). The operational principle of the communication with such a cloud-based secure element is similar to that of a relay channel. Thus, these systems will only work with relaxed timing requirements.

However, newer concepts for cloud-based secure element concepts use tokenization techniques to create temporary payment credentials stored within the mobile device memory. Hence, they cache the data necessary to perform transactions on the mobile device ("offline") and, therefore, do not require such relaxed timing requirements.

6.4.1.3 Other Timing Constraints

Typical limits for contactless transactions in transport ticketing and payment are between 300 and 500 ms (cf. [31]). These limits apply to overall transactions, which, typically, consist of multiple command-response pairs. However, these limits are meant to maintain user experience and are usually not enforced as maximum timeouts by contactless reader devices. For instance, Francis et al. [6] measured that a reader for machine-readable travel documents which they used for relay attacks would interrupt transactions only if they were not completed after 5.2 s.

6.4.2 Building a Card Emulator

There are several different options when building a proxy card emulation device. These options range from building new card emulator hardware from scratch to using various kinds of existing hardware. The various choices have different advantages and disadvantages that result into different design and implementation costs and into different levels of emulation capabilities.

6.4.2.1 Building a New Device from Scratch

This method gives full control over the whole design process. The card emulator can be put into any inconspicuous looking shape. The whole RFID protocol stack can be controlled starting from the lowest layer. Thus, parameters of all protocol layers can be adapted. For instance, the unique identifier (UID) that is used during the anti-collision sequence and the ATS could be freely chosen. Also, low-level protocol timing can be influenced by setting FWT values and by handling the WTX mechanism. However, building a new emulator device from scratch also involves the highest design costs and effort.

6.4.2.2 Using a Ready-Made RFID Card Emulation Device

Dedicated card emulation hardware for ISO/IEC 14443 can be bought on the Internet. For example, there exist the IAIK HF DemoTag [32] and the Proxmark [21]. Both provide a hardware platform and a rudimentary software stack for card emulation. With this choice, it is still possible to control the whole RFID protocol stack. However, the ready-made devices cannot easily be fit into any desired shape.

6.4.2.3 Using an NFC Reader Device

Some NFC reader devices (e.g. the ACS ACR 122U) can not only read contactless smartcards and communicate in peer-to-peer mode, but they can also be put into software card emulation mode. In this mode, the device waits for APDU commands from an external reader and forwards them to the computer. The computer then generates a response that is returned to the reader. The lower protocol layers are handled automatically by the reader firmware. For instance, NXP's PN532 NFC controller that is embedded into the ACR 122U even performs automatic waiting time extension.

One disadvantage of this approach is that many NFC chipsets do not allow the user to freely choose all parameters for the protocol layers below APDUs. For example, with NXP's PN532 the UID for the anti-collision procedure must always start with 0x08, which denotes a random ID. Also, the ACR 122U can only emulate the ISO/IEC 14443 Type A communication protocol. Moreover, the ready-made NFC reader cannot easily be fit into any desired (inconspicuous) shape.

6.4.2.4 Using an NFC-Enabled Mobile Phone

Yet another alternative is the use of NFC-enabled mobile phones as software card emulation devices. While BlackBerry mobile phones were the first phones that contained an API [23] for software card emulation, other NFC devices could be adapted to support software card emulation as well. User-contributed patches [37, 38] to the CyanogenMod 9.1 aftermarket Android firmware show that open mobile phone platforms can easily be extended to enable this type of card emulation. Starting with Android 4.4, software card emulation (under the term *host-based card emulation*, HCE [1]) has been integrated into the official Android platform and is supported on a broad range of Android NFC devices.

Using a mobile phone as card emulation device has several advantages (cf. [26]). First, the mobile phone already has the form-factor that is expected for NFC contactless transactions. In other words, it is a mobile phone just as the device that actually carries the secure element and that is expected to be used at e.g. the point-of-sale. Second, the mobile phone has network interfaces that match the network interfaces of the device-under-attack. Thus, it can be easily connected to the relay software. Third, the mobile phone has all the processing capabilities to transfer APDU commands between its network interface and its card emulation hardware. Consequently, no extra hardware, like a PC, is required.

On BlackBerry mobile phones, software card emulation is possible with both, ISO/IEC 14443 Type A and Type B communication protocols. On Android mobile phones, support for card emulation depends on the NFC chipset and the device implementation. Devices either support ISO/IEC 14443 Type A or Type B. Both, Android and BlackBerry mobile phones with card emulation support only allow emulation of protocols on top of the ISO/IEC 14443-4 block transmission protocol. Android even requires the use of APDUs and the ISO/IEC 7816-4 application selection mechanism in order to route the communication to HCE apps. While this is sufficient for the relay scenario described in this thesis, it is not possible to influence or handle lower protocol layers (e.g. to implement proprietary protocols on top of ISO/IEC 14443-3) using the APIs available on these platforms. Also, none of these platforms allow low-level parameters, like the anti-collision identifier (UID), to be freely chosen. Woolley [36] explains that, for the BlackBerry platform, this was an intentional design decision due to security concerns.

6.4.3 Prototype Implementation of the Relay System

An initial prototype has been developed to proof the concept of the software-based relay attack. The prototype consists of the following parts:

- a Samsung Nexus S with Android 2.3.4,
- an Android app (relay software) that accesses the hidden secure element API (com.android.nfc_extras) and relays commands over a TCP (Transmission Control Protocol) socket,
- a Python script (card emulation software) that controls the card emulation hardware and relays commands over a TCP socket, and
- an ACS ACR 122U NFC reader in software card emulation mode.[1]

---

[1] At the time this research was conducted, software card emulation (as described in Sect. 6.4.2.4) was not yet available on Android phones.

6.4.3.1 The Relay App

The relay app, a purely Java-based Android app, is a simple TCP client that maintains a persistent TCP connection to a remote server (the card emulator). The IP (Internet Protocol) address of the remote server can be set in a simple user interface (Fig. 6.8). Using the mobile phone as the client and the card emulator as the server has the advantage that connections are even possible when the mobile network operator blocks all incoming connections to the mobile phone. Once connected, the server can send commands to the client. The most important commands are:

- connect to the secure element,
- transmit APDU and receive response,
- disconnect from the secure element, and
- terminate the relay app.

When the card emulator requests access to the secure element, a connection is established through the hidden Android secure element API (class `NfcExecutionEnvironment` in `com.android.nfc_extras`). The relay app listens for command APDUs on the TCP socket and forwards them to the secure element. The response APDUs from the secure element are transmitted back through the TCP socket to the card emulator. The following listing gives an overview of the functionality of the relay app in pseudo-code:

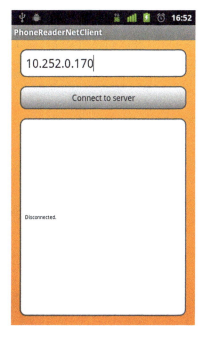

Fig. 6.8 Android relay app

```
1   TCPClientSocket->ConnectToServer(IPAddress);
2
3   loop {
4       command := TCPClientSocket->WaitForCommand();
5       switch command {
6           case ConnectSecureElement:
7               SecureElement->open();
8
9           case SendCommandAPDU:
10              responseAPDU := SecureElement->transceive(
11                      command->commandAPDU);
12              TCPClientSocket->SendResponseAPDU(
13                      responseAPDU);
14
15          case DisconnectSecureElement:
16              SecureElement->close();
17
18          case TerminateRelayApp:
19              exit loop;
20      }
21  }
22
23  TCPClientSocket->DisconnectFromServer();
```

6.4.3.2 The Card Emulator

The card emulator (Fig. 6.9) has been built from an ACS ACR 122U NFC reader and a notebook computer running a card emulation server application. The ACR 122U supports software card emulation mode and is available for about EUR 50 (including taxes) from GoToTags [10].

The card emulation software, a Python application, is based on the nfcpy [33] project. The nfcpy project has been modified to enable software card emulation for the PN532 NFC controller in the ACR 122U. The card emulation software contains

Fig. 6.9 Card emulator made from a notebook and an ACS ACR 122U NFC reader (*Source* [25])

6.4 Viability of the Software-Based Relay Attack

a TCP server that listens for incoming connections from the relay app. Once a TCP connection has been established, the card emulation server requests access to the secure element through the relay app:

```
1  TCPClientSocket := TCPServer->WaitForClient(...);
2  TCPClientSocket->ConnectSecureElement();
```

Then, it puts the PN532 into card emulation mode:

```
3  PN532->WriteRegister(
4          CIU_TxMode,
5          TxCRCEn | TxSpeed := 106 kbps |
6          TxFraming := ISO/IEC 14443A);
7  PN532->WriteRegister(
8          CIU_RxMode,
9          RxCRCEn | RxSpeed := 106 kbps |
10         RxFraming := ISO/IEC 14443A);
11 PN532->WriteRegister(CIU_TxAuto, InitialRFOn);
12
13 PN532->SetParameters(fISO14443-4_PICC | fAutomaticRATS);
14
15 PN532->TgInitAsTarget(
16         PICCOnly | PassiveOnly,
17         MifareParams := {
18                 SENS_RES := { 0x00, 0x04} |
19                 NFCID1t := {0x76, 0x82, 0x4F} |
20                 SEL_RES := 0x20 },
21         FelicaParams := { 0x00, ..., 0x00 },
22         NFCID3t := { 0x00, ..., 0x00 },
23         GeneralBytes := { },
24         HistoricalBytes := {
25                 0x80, 0x31, 0x80, 0x66, 0xB0, 0x84,
26                 0x0C, 0x01, 0x6E, 0x01, 0x83, 0x00,
27                 0x90, 0x00 });
```

The emulator then waits for commands from a contactless smartcard reader:

```
28 loop {
29     commandAPDU := PN532->TgGetData();
30     if (commandAPDU == ERROR) { exit loop; }
31     TCPClientSocket->SendCommandAPDU(commandAPDU);
```

Upon reception of a command APDU, the card emulation server forwards it through the TCP socket to the relay app. As soon as a response is received on the TCP socket, the response is returned to the smartcard reader:

```
32     responseAPDU := TCPClientSocket->WaitForResponseAPDU();
33     PN532->TgSetData(responseAPDU);
34 }
```

The card emulator then waits for the next command APDU. This cycle continues until the ACR 122U leaves the range of the smartcard reader (i.e. until an RF field is no longer detected). Then, the card emulator instructs the relay app to close the connection to the secure element:

```
35 TCPClientSocket->DisconnectSecureElement();
```

Figure 6.10 shows a sequence diagram of the interaction between smartcard reader, card emulation server, relay app and secure element.

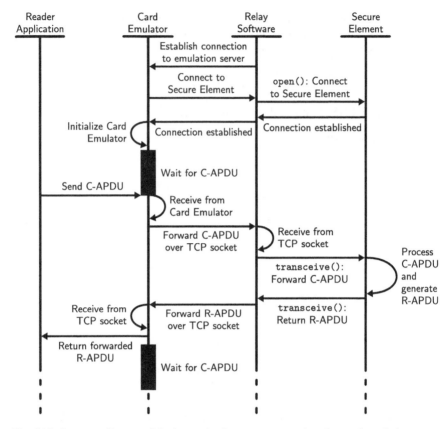

Fig. 6.10 Sequence diagram of the interaction between smartcard reader, card emulation server, relay app and secure element (*Source* [28])

6.4.4 Test Setup for Measurement of Communication Delays

Four different scenarios are compared to verify the feasibility of the relay system:

1. The secure element is accessed directly in external mode with an external smartcard reader. An app on the mobile phone activates the secure element for external access.
2. The secure element is accessed directly in internal mode with an app on the phone.
3. The secure element is accessed through the relay system using a direct Wi-Fi link between the phone and the card emulator.
4. The secure element is accessed through the relay system using the mobile phone network and the Internet between the phone and the card emulator.

For each scenario, the time between sending a command and receiving a response is measured. The first two scenarios do not use a relay channel but provide reference values to compare relay and non-relay cases instead.

6.4 Viability of the Software-Based Relay Attack

By the time of the measurements, there were no secure element-based applications (e.g. a credit card or an access control applet) available for the Nexus S in Austria. Also, we do not have access to the secret keys that are required for GlobalPlatform card management. Thus, we were unable to install a custom application onto the secure element. As a consequence, a different approach had to be found to measure communication delays with the secure element. Therefore, we chose to communicate with the only applet that was known to be available on the blank, GlobalPlatform-compliant secure element: the GlobalPlatform card manager application for the issuer security domain.

6.4.4.1 Customized Reader

The customized smartcard reader consists of a Java SE application and an HID OMNIKEY 5321 USB contactless reader connected through PC/SC (Personal Computer/Smart Card Interface). The reader application interacts with the GlobalPlatform card manager by exchanging a sequence of three APDU commands:

1. SELECT the issuer security domain by its AID:
   00 A4 0400 08 **A000000003000000** (13 bytes)
   A file control information template is expected as response:
   ```
   6F 65
      84 08 A000000003000000
      A5 59
         9F65 01 FF
         9F6E 06 479100783300
         73 4A
            06 07 2A864886FC6B01
            60 0C 06 0A 2A864886FC6B02020101
            63 09 06 07 2A864886FC6B03
            64 0B 06 09 2A864886FC6B040215
            65 0B 06 09 2B8510864864020103
            66 0C 06 0A 2B060104012A026E0102
   9000 (105 bytes)
   ```
2. GET_DATA for data object 0x65:
   00 CA **0065** 00 (5 bytes)
   The expected response is a "reference data not found" error code:
   6A88 (2 bytes)
3. GET_DATA for data object 0x66:
   00 CA **0066** 00 (5 bytes)
   The expected response is the card data:
   ```
   73 4A
      06 07 2A864886FC6B01
      60 0C 06 0A 2A864886FC6B02020101
      63 09 06 07 2A864886FC6B03
   ```

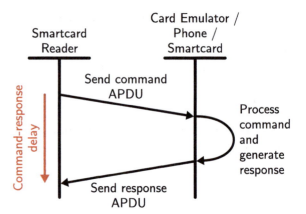

Fig. 6.11 Sequence diagram that shows the measured command-response delay

```
    64 0B 06 09 2A864886FC6B040215
    65 0B 06 09 2B8510864864020103
    66 0C 06 0A 2B060104012A026E0102
9000 (78 bytes)
```

As soon as a smartcard is detected in the RF field of the reader, the application connects to that card, sends the three APDU commands and disconnects. The reader application measures the delay in milliseconds between sending the command and receiving the response for each APDU command-response pair (cf. Fig. 6.11):

```
1  long timeStart = System.currentTimeMillis();
2  byte[] responseAPDU = smartcard.exchangeData(commandAPDU);
3  long timeEnd = System.currentTimeMillis();
4  long delay = timeEnd - timeStart;
```

6.4.4.2 Reader App for Reference Measurement

The customized reader app can be used in all but one scenario. For scenario 2 the command-response delay must be measured directly in an app on the mobile phone. Therefore, an Android app with equivalent functionality to the external reader app has been developed. The app opens a connection to the secure element, exchanges the three APDUs and closes the connection again. For each APDU, the app measures the command-response delay:

6.4 Viability of the Software-Based Relay Attack

```
1  long timeStart = System.currentTimeMillis();
2  byte[] responseAPDU = securelement.transceive(commandAPDU);
3  long timeEnd = System.currentTimeMillis();
4  long delay = timeEnd - timeStart;
```

6.4.4.3 Automated Test System

Figure 6.12 shows the test system for measurements of the communication delays. In order to get a significant result, several thousand communication cycles between the secure element and the smartcard reader had to be recorded. Between each cycle the card emulator (or the phone) had to be isolated from the smartcard reader to maintain identical conditions for each measurement. Tests with scenario 3 showed that the average command-response delay does not significantly deviate for more

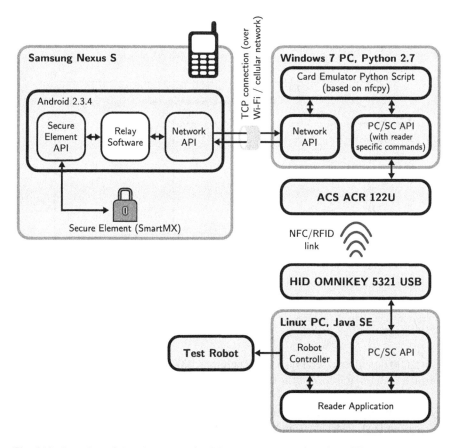

Fig. 6.12 Overview of the relay system for delay measurements (based on [28])

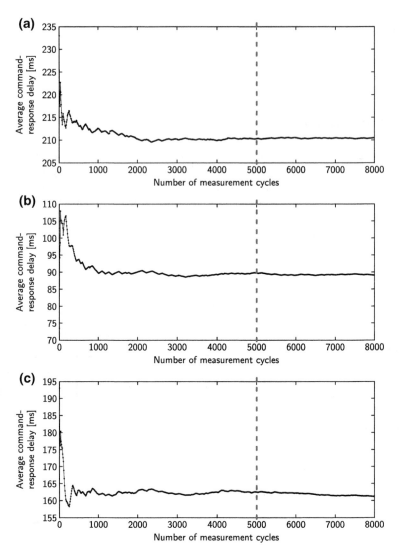

Fig. 6.13 Variation of the average command-response delay over the number of measurement cycles: **a** SELECT, **b** GET_DATA for data object 0x65, **c** GET_DATA for data object 0x66

than 5,000 repetitions (see Fig. 6.13). Therefore, 5,000 repetitions have been chosen for each scenario to account for variations of the communication channels.

As 15,000 repetitions (5,000 for each of the scenarios 1, 3 and 4) would take a long time if they are performed manually, an automated test setup has been created. The customized reader is extended with a test robot. The test robot moves the card emulator (or the phone) into the RF field of the reader. After the reader software completed the exchange of the three APDU command-response pairs, the test robot

6.4 Viability of the Software-Based Relay Attack

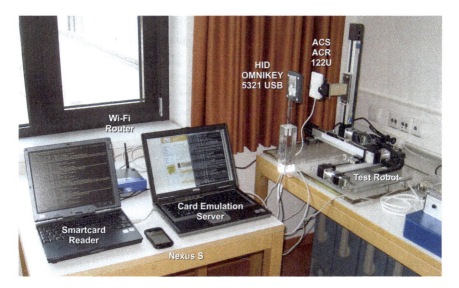

Fig. 6.14 Test setup for measurements of the communication delay

removes the card emulator (or the phone) from the RF field of the reader. This procedure is repeated for each of the 5,000 cycles. The complete test setup is shown in Fig. 6.14.

6.4.5 Measurement Results

The four scenarios are compared based on measurements performed with the test system described above.

6.4.5.1 Measured Paths

Figure 6.15 shows the paths that are used for measurement of the communication channel in each of the four scenarios:

- *Scenario 1*: The external interface of the secure element is accessed directly with the reader system. The command-response delay is measured in the reader software.
- *Scenario 2*: The internal interface of the secure element is accessed with an app on the phone. The command-response delay is measured by the app.
- *Scenario 3*: The internal interface of the secure element is accessed by the relay app. The relay app is connected to a Wi-Fi router through the Wi-Fi interface of the mobile phone. The card emulation server is connected to the Wi-Fi router through

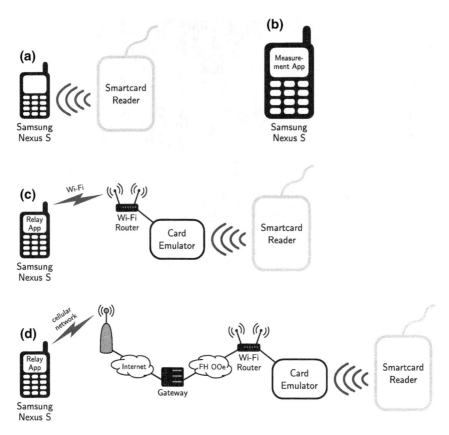

Fig. 6.15 Communication paths for command-response delay measurement: **a** scenario 1, **b** scenario 2, **c** scenario 3, **d** scenario 4

a direct cabled interface. The card emulator is accessed by the reader system. The command-response delay is measured in the reader software.

- *Scenario 4*: The internal interface of the secure element is accessed by the relay app. The relay app is connected to the Internet through the cellular network interface of the mobile phone. The relay app accesses the card emulation server through the public IP address 193.170.124.15. This IP address is assigned to the WAN port of the Wi-Fi router and routed through the FH Oberösterreich corporate network.[2] The card emulation server is connected to the Wi-Fi router through a direct cabled interface. The card emulator is accessed by the reader system. The command-response delay is measured in the reader software.

---

[2] Unfortunately, the mobile phone operator (A1 Telekom Austria) blocks the packets necessary to perform a traceroute analysis. Therefore, the exact routing between the mobile phone and the Wi-Fi router could not be determined.

6.4 Viability of the Software-Based Relay Attack

6.4.5.2 Results for the SELECT Command

Figure 6.16 shows the histograms of the command-response delay for the APDU "SELECT issuer security domain by AID" for 5,000 repetitions. The histograms are divided into 200 bins. Each bin has a width of 50 ms. The last bin also contains all measurements above 10,000 ms. Figure 6.16a is zoomed from 0 to 50 ms with 1-ms-bins. Figure 6.16a is zoomed from 0 to 150 ms with 5-ms-bins. Figure 6.16c is zoomed from 0 to 500 ms with 5-ms-bins.

The delay for scenario 1 centers on about 30 ms. On-device access to the secure element (scenario 2) already takes significantly longer (50–80 ms). The delay over the Wi-Fi link (scenario 3) ranges from 190 to 260 ms. Thus, the Wi-Fi relay link adds a delay in the range of 110 and 210 ms. For scenario 4, the delays start at about 200 ms and have a significant peak around 300 ms. About 44 % of the measured delays are below 1 s, about 80 % are below 4 s, and about 97 % are below 10 s.

6.4.5.3 Results for the GET_DATA Command for Data Object 0x65

Figure 6.17 shows the histograms of the command-response delay for the APDU "GET_DATA for data object 0x65" for 5,000 repetitions. The histograms are divided into 200 bins. Each bin has a width of 10 ms. The last bin also contains all measurements above 2,000 ms. Figure 6.17a is zoomed from 0 to 50 ms with 1-ms-bins. Figure 6.17b is zoomed from 0 to 150 ms with 5-ms-bins. Figure 6.17c is zoomed from 0 to 500 ms with 5-ms-bins.

The delay for scenario 1 centers on about 13 ms. On-device access to the secure element (scenario 2) already takes significantly longer (20–40 ms). The delay over the Wi-Fi link (scenario 3) ranges from 55 to 135 ms. Thus, the Wi-Fi relay link adds a delay in the range of 15 and 115 ms. For scenario 44, the delays start at about 120 ms and have a significant peak around 170 ms.

6.4.5.4 Results for the GET_DATA Command for Data Object 0x66

Figure 6.18 shows the histograms of the command-response delay for the APDU "GET_DATA for data object 0x66" for 5,000 repetitions. The histograms are divided into 200 bins. Each bin has a width of 10 ms. The last bin also contains all measurements above 2,000 ms. Figure 6.18a is zoomed from 0 to 50 ms with 1-ms-bins. Figure 6.18b is zoomed from 0 to 150 ms with 5-ms-bins. Figure 6.18c is zoomed from 0 to 500 ms with 5-ms-bins.

The delay for scenario 1 centers on about 19 ms. On-device access to the secure element (scenario 2) already takes significantly longer (35–60 ms). The delay over the Wi-Fi link (scenario 3) ranges from 100 to 220 ms. Thus, the Wi-Fi relay link adds a delay in the range of 40 and 185 ms. For scenario 4, the delays start at about 160 ms and have a significant peak around 240 ms.

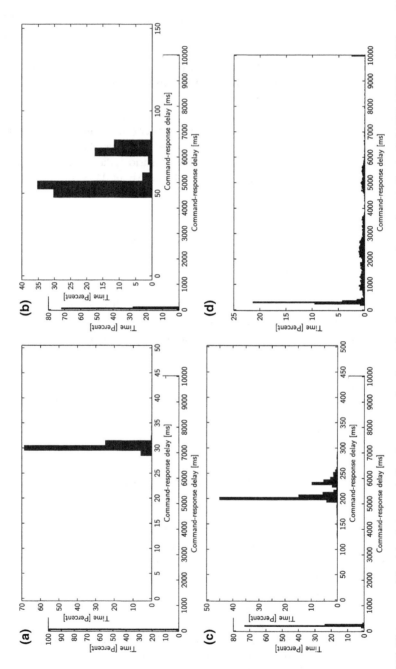

Fig. 6.16 Histograms of the delay between sending the command and receiving the response at the reader side for the SELECT APDU at 5,000 repetitions: **a** scenario 1, **b** scenario 2, **c** scenario 3, **d** scenario 4 (*Source* [28])

6.4 Viability of the Software-Based Relay Attack

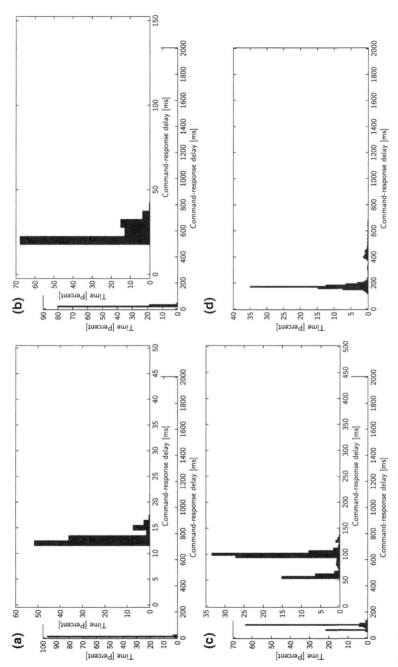

Fig. 6.17 Histograms of the delay between sending the command and receiving the response at the reader side for the GET_DATA APDU for data object 0x65 at 5,000 repetitions: **a** scenario 1, **b** scenario 2, **c** scenario 3, **d** scenario 4

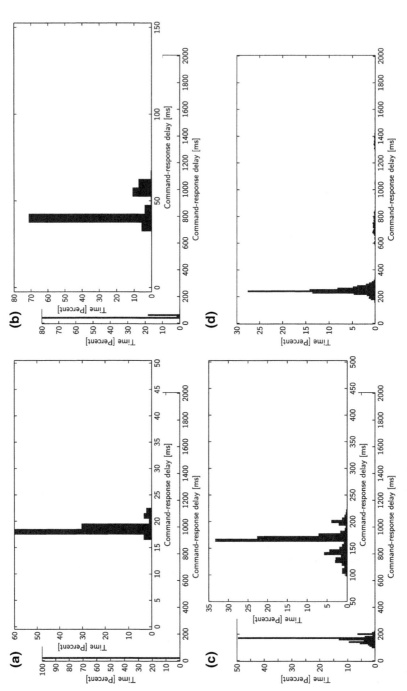

Fig. 6.18 Histograms of the delay between sending the command and receiving the response at the reader side for the GET_DATA APDU for data object 0x66 at 5,000 repetitions: **a** scenario 1, **b** scenario 2, **c** scenario 3, **d** scenario 4

6.4.5.5 Comparison of the Results

The Wi-Fi link increased delays around 100–200 ms per APDU sequence. The Internet link, on average, added about 200 ms of delay to each APDU sequence. For a transaction that consists of multiple APDUs, the relay channel may add several seconds to the total transaction time.

However, the measurements revealed that not only the relay channel adds to the overall command-response delay. On the Nexus S, access to the secure element in internal mode is also significantly slower (up to 50 ms) than access in external mode.

The limits for contactless transactions that are usually expected in transport ticketing and payment (300–500 ms) will most likely not be met for relayed transactions. Nevertheless, typical point-of-sale terminals would accept a transaction even if it takes up to 20 s. Thus, a relayed transaction would be accepted by such terminals even if it consists of several APDU command-response pairs. As contactless transactions (especially with mobile phones) are quite new and users are still not used to them, we assume that even long delays (in the order of 20–30 s per transaction) will not raise suspicions to point-of-sale personnel.

6.5 Possible Solutions

Several countermeasures against relay attacks have been identified in publications about relay attacks on contactless smartcards [11, 12, 19]:

1. The contactless RF interface can be shielded with a Faraday cage when it is not in use.
2. A smartcard could contain additional circuitry for physical activation and deactivation.
3. Additional passwords or PIN codes could be used for two-factor authentication.
4. Distance bounding protocols can be used on fast channels to determine the actual distance between the smartcard and the reader.

An analysis of these proposed methods with regard to the software-based relay attack reveals that not all of them can be applied to internal card emulation mode:

1. Shielding with a Faraday cage is only possible for external mode. The equivalent of shielding for internal mode is an API access control policy. However, access control mechanisms have been found to have weaknesses on many platforms (cf. Sect. 6.2). Nevertheless, the adoption of trusted computing concepts for mobile phone systems could potentially improve the security of the interaction between the application processor and the secure element. For instance, the secure element could recognize if the application processor is in a trusted state.
2. Physical activation and deactivation of the secure element (e.g. by means of a hardware button on the mobile device) would significantly hinder over-the-air card management. Thus, it seems to be impractical to implement physical activation and deactivation.

3. Two-factor authentication can reliably hinder relay attacks as long as PINs are not known by the attacker. While an attacker would still be able to relay the communication, the applications on the secure element would be unusable without knowledge of the PINs. Nevertheless, if PINs are entered on the mobile phone, an attacker with sufficient privileges might be able to monitor and reuse previously entered PIN codes.
4. Distance bounding protocols would require an additional fast communication channel (e.g. ultra-wideband [12]). Such a channel is neither defined in current NFC standards nor available in current NFC chipsets.

Section 6.4.1 revealed that protocols involved in contactless transactions have insufficient timing constraints to provide adequate protection against relay attacks. Still, some (potentially unreliable) timing-based countermeasures are conceivable:

- Contactless readers—particularly EMV credit card terminals—could enforce shorter timeouts. For instance, the EMV specifications define rather tight benchmark targets for credit card transactions that must be achieved by both the terminal and the card (500 ms per contactless payment transaction [2]). Thus, a POS terminal could safely interrupt and disregard transactions taking longer than this timeout if the credit card complies with the EMV specifications. Nonetheless, this measure would not prevent relay over shorter distances and fast communication channels. Also, such tight timeouts would prevent cloud-based EMV applications where the secure credit card is stored in the cloud and accessed over the Internet (cf. [26]) as these applications would have delays comparable to the malicious relay scenario.
- The processing time of a command on a smartcard is usually similar for each invocation of the command with the same side conditions. Therefore, significant deviations of command-response delays in comparison to previous transactions may indicate a change in the communication channel and, thus, may be a symptom of a relay attack. As a consequence, readers could base timeouts on a history of measured command-response delays from previous transactions with that card.
- Similarly, timeouts could be established for specific commands that are known to have a short processing time on the smartcard.

Another option would be to disable internal mode communication for certain applications or commands. Many current secure element microchips provide instruments to distinguish between external mode communication and internal mode communication. Based on this information, applets as a whole can be restricted to using a specific interface (either internal mode or external mode). Moreover, applets themselves can use this information to decide on a per-APDU basis whether a specific command should be allowed in internal mode or external mode. However, such restrictions are not always possible. For example, over-the-air card management must be possible through internal mode.[3] Therefore, the denial-of-service attack described in Sect. 6.3.1 may still be applicable. Also for credit card payment applications, a significant advantage of the secure element would be that these applications could

---

[3] The UICC is an exception to this as over-the-air card management of the UICC may be possible directly without access from the application processor.

6.5 Possible Solutions

be used for secure (card-present) transactions from within the mobile phone (e.g. through the mobile phone web browser). Therefore, if access to credit card applications is limited to external mode in favor of preventing software-based relay attacks, these card-present transactions would not be possible.

References

1. Android Open Source Project: Android developers—host-based card emulation. https://developer.android.com/guide/topics/connectivity/nfc/hce.html (2014). Accessed Dec 2014
2. EMVCo: EMV Contactless Specifications for Payment Systems—Book A: Architecture and General Requirements. Version 2.1 (2011)
3. Forristal, J.: Android fake ID vulnerability. Talk at BlackHat US. Las Vegas, NV, USA. https://www.blackhat.com/docs/us-14/materials/us-14-Forristal-Android-FakeID-Vulnerability-Walkthrough.pdf (2014)
4. Forum Nokia: Nokia 6131 NFC SDK: user's guide. Version 1.1 (2007)
5. Francis, L., Hancke, G.P., Mayes, K.E., Markantonakis, K.: Practical NFC peer-to-peer relay attack using mobile phones. In: Radio Frequency Identification: Security and Privacy Issues, LNCS, vol. 6370/2010, pp. 35–49. Springer, Berlin (2010). doi:10.1007/978-3-642-16822-2_4
6. Francis, L., Hancke, G.P., Mayes, K.E., Markantonakis, K.: Practical relay attack on contactless transactions by using NFC mobile phones. Cryptology ePrint Archive, Report 2011/618. http://eprint.iacr.org/2011/618 (2011)
7. Giesecke & Devrient: seek-for-android—Secure Element Evaluation Kit for the Android Platform—the 'SmartCard API'. https://code.google.com/p/seek-for-android/ (2012). Accessed Sept 2012
8. GlobalPlatform: Card Specification. Version 2.2.1 (2011)
9. GlobalPlatform: Secure Element Access Control. Version 1.0 (2012)
10. GoToTags: GoToTags—the nfc superstore. https://www.buynfctags.com/ (2012). Accessed Sept 2012
11. Hancke, G.P.: A practical relay attack on ISO 14443 proximity cards. http://www.rfidblog.org.uk/hancke-rfidrelay.pdf (2005). Accessed Sept 2011
12. Hancke, G.P., Mayes, K.E., Markantonakis, K.: Confidence in smart token proximity: relay attacks revisited. Comput. Secur. **28**(7), 615–627 (2009). doi:10.1016/j.cose.2009.06.001
13. International Organization for Standardization: ISO/IEC 14443-4: Identification cards—Contactless integrated circuit cards—Proximity cards—Part 4: Transmission protocol (2008)
14. International Organization for Standardization: ISO/IEC 14443-2: Identification cards—Contactless integrated circuit cards—Proximity cards—Part 2: Radio frequency power and signal interface (2010)
15. International Organization for Standardization: ISO/IEC 14443-3: Identification cards—Contactless integrated circuit cards—Proximity cards—Part 3: Initialization and anticollision (2011)
16. Issovits, W., Hutter, M.: Weaknesses of the ISO/IEC 14443 protocol regarding relay attacks. In: Proceedings of the IEEE International Conference on RFID-Technologies and Applications (RFID-TA), pp. 335–342. IEEE, Sitges, Spain (2011). doi:10.1109/RFID-TA.2011.6068658
17. Java Community Process: JSR 177: Security and Trust Services API (SATSA). Version 1.0.1 (2007)

18. Java Community Process: JSR 257: Contactless Communication API. Version 1.1 (2009)
19. Kfir, Z., Wool, A.: Picking virtual pockets using relay attacks on contactless smartcard. In: Proceedings of the First International Conference on Security and Privacy for Emerging Areas in Communications Networks (SecureComm 2005), pp. 47–58. IEEE, Athens, Greece (2005). doi:10.1109/SECURECOMM.2005.32
20. MasterCard Worldwide: Transaction optimization for PayPass—M/Chip terminals. Number 002582, version 2–0 (2008)
21. proxmark.org: Proxmark—a radio frequency identification tool. http://www.proxmark.org/ (2012). Accessed Sept 2012
22. Research In Motion: BlackBerry code signing keys—BlackBerry keys order form. https://www.blackberry.com/SignedKeys/ (2012). Accessed Sept 2012
23. Research In Motion: Blackberry API 7.0.0: package net.rim.device.api.io.nfc.emulation. http://www.blackberry.com/developers/docs/7.0.0api/net/rim/device/api/io/nfc/emulation/package-summary.html (2011). Accessed Sept 2012
24. Research In Motion: Blackberry API 7.0.0: package net.rim.device.api.io.nfc.se. http://www.blackberry.com/developers/docs/7.0.0api/net/rim/device/api/io/nfc/se/package-summary.html (2011). Accessed Sept 2012
25. Roland, M.: Applying recent secure element relay attack scenarios to the real world: Google Wallet relay attack. Computing Research Repository (CoRR), arXiv:1209.0875 [cs.CR]. http://arxiv.org/abs/1209.0875 (2012)
26. Roland, M.: Software card emulation in NFC-enabled mobile phones: great advantage or security nightmare? In: 4th International Workshop on Security and Privacy in Spontaneous Interaction and Mobile Phone Use. Newcastle, UK. http://www.medien.ifi.lmu.de/iwssi2012/papers/iwssi-spmu2012-roland.pdf (2012)
27. Roland, M., Langer, J., Scharinger, J.: Practical attack scenarios on secure element-enabled mobile devices. In: Proceedings of the Fourth International Workshop on Near Field Communication (NFC 2012), pp. 19–24. IEEE, Helsinki, Finland (2012). doi:10.1109/NFC.2012.10
28. Roland, M., Langer, J., Scharinger, J.: Relay attacks on secure element-enabled mobile devices: virtual pickpocketing revisited. In: Information Security and Privacy Research, IFIP AICT, vol. 376/2012, pp. 1–12. Springer, Heraklion, Creete, Greece (2012). doi:10.1007/978-3-642-30436-1_1
29. Roland, M., Langer, J., Scharinger, J.: Applying relay attacks to Google Wallet. In: Proceedings of the Fifth International Workshop on Near Field Communication (NFC 2013). IEEE, Zurich, Switzerland (2013). doi:10.1109/NFC.2013.6482441
30. SIMalliance: Open Mobile API specification. Version 2.03 (2012)
31. Smart Card Alliance Transportation Council: Transit and contactless open payments: an emerging approach for fare collection. White paper. http://www.smartcardalliance.org/resources/pdf/Open_Payments_WP_110811.pdf (2011)
32. Stiftung Secure Information and Communication Technologies: HF RFID Demo Tag. http://jce.iaik.tugraz.at/Products/RFID-Components/HF-RFID-Demo-Tag (2012). Accessed Sept 2012
33. Tiedemann, S.: nfcpy—Python module for near field communication. https://launchpad.net/nfcpy (2012). Accessed Sept 2012
34. Urien, P., Piramuthu, S.: Securing NFC mobile services with cloud of secure elements (CoSE). In: Mobile Computing, Applications, and Services, LNICST, vol. 130/2014, pp. 322–331. Springer, Paris, France (2013). doi:10.1007/978-3-319-05452-0_30
35. Urien, P., Piramuthu, S.: Towards a secure cloud of secure elements concepts and experiments with NFC mobiles. In: Proceedings of the 2013 International Conference on Collaboration Technologies and Systems (CTS), pp. 166–173. IEEE, San Diego, CA, USA (2013). doi:10.1109/CTS.2013.6567224

36. Woolley, M.: Response to thread "UID for NFC Mifare Tag emulation" on BlackBerry support community forums. http://supportforums.blackberry.com/t5/Java-Development/UID-for-NFC-Mifare-Tag-emulation/m-p/1575809 (2012)
37. Yeager, D.: Added NFC Reader support for two new tag types: ISO PCD type A and ISO PCD type B. https://github.com/CyanogenMod/android_packages_apps_Nfc/commit/d41edfd794d4d0fedd91d561114308f0d5f83878 (2012)
38. Yeager, D.: Added NFC Reader support for two new tag types: ISO PCD type A and ISO PCD type B. https://github.com/CyanogenMod/android_external_libnfc-nxp/commit/34f13082c2e78d1770e98b4ed61f446beeb03d88 (2012)
39. YES-wallet: YES-wallet—cloud-based NFC contactless mobile wallet.http://www.yes-wallet.com/ (2012). Accessed Sept 2012

Chapter 7
Software-Based Relay Attacks on Existing Applications

The software-based relay attack has been applied to *Google Wallet* [7] to verify its feasibility in existing payment systems. Google Wallet has been chosen for several reasons:

- Google Wallet was made available to users outside the US through XDA Developers forum[1] in late 2011 using some hacks to bypass mobile phone operator and location restrictions.
- Google Wallet is already in use by many users. Google Play Store[2] listed more than 500,000 installations in early 2012. By the end of 2012, Google Wallet had over 1,000,000 installations. Meanwhile, it has over 10,000,000 installations.
- The wallet contains a credit card application that is based on EMV payment standards (specifically on *MasterCard PayPass* using the EMV Mag-Stripe mode) and can be used with any point-of-sale terminal that supports PayPass contactless credit card transactions.
- The Android source code is publicly available. Thus, it was fairly easy to explore the Near Field Communication (NFC) software stack of Android devices and the hidden secure element application programming interface (secure element API, cf. Sect. 6.2.4.1).
- Google Wallet is known to be installed by many users on rooted devices (mainly to circumvent operator and location restrictions). This means that the security measures of the operating system are already weakened/bypassed on those phones.
- For non-rooted devices, there either already exist privilege escalation exploits or it is assumed that such exploits will appear soon (cf. [19]). Additionally, once an exploit is found/known, it can take several months until devices in the field are patched [1].

---

[1] http://forum.xda-developers.com/showthread.php?t=1365360.
[2] https://play.google.com/store/apps/details?id=com.google.android.apps.walletnfcrel.

7.1 Google Wallet

Google Wallet is a container for payment cards, gift cards, reward cards and special offers. It consists of an Android app with a user interface and Java Card applets on the secure element.[3] The user interface is used to protect the wallet with a PIN code, to manage the payment, gift and reward cards, to select the currently active card, to find specific offers and to view the transaction history. The secure element is used to store sensitive information of the payment, gift and reward cards and to interact with existing point-of-sale (POS) reader infrastructures. The analysis and attack described in this thesis have been performed with version 1.1-R52v7 of the Google Wallet app and the secure element applets installed in February 2012.

7.1.1 Preparing for an In-depth Analysis

For this analysis of Google Wallet, the most interesting part is the communication with the secure element. Therefore, debug output has been added to the Android secure element API (com.android.nfc_extras) in order to trace the interaction with the API and the communication between the Google Wallet app and the secure element. As the secure element API is encapsulated in a separate library (JAR file), it was possible to modify and re-compile the library source code. Our modifications output messages to the Android debug log for the following operations:

- When an app enables or disables the external mode of a secure element.
- When an app looks up the current state of the external mode activation of a secure element.
- When an app retrieves an instance of the NfcExecutionEnvironment class for internal mode communication.
- When an app opens or closes a connection (internal mode) to the secure element.
- When an app exchanges APDUs (application protocol data units) with the secure element. In this case, the debug output also contains the exchanged APDUs.

The log messages reveal the method, the method parameters and the return values. Appendix B lists the modified source code of com.android.nfc_extras. After re-compilation, the library JAR file on the device (/system/framework/com.android.nfc_extras.jar) has been replaced with the modified version.

---

[3] Since host-based card emulation was introduced to Android, recent versions of the wallet can use a cloud-based secure element in combination with host-based card emulation instead of an on-device secure element.

7.1.2 Static Structure

Upon first start, the Google Wallet app initializes the secure element and installs a PIN code that is necessary for using the app user interface. During initialization several applets are installed and personalized on the secure element using GlobalPlatform card management (cf. [4]). Specifically, the Google Wallet app creates a connection between the secure element and a remote server. Over that connection, the remote server establishes a secure channel based on the secure channel protocol *SCP02* to the secure element. The remote server then performs the card management through this authenticated and partly encrypted channel.

While sensitive information—like applets and data that is personalized into these applets—is encrypted, responses from the secure element to the remote server are sent without encryption. Thus, it was possible to extract some information from these responses. One such information is the list of executable load-files present on the secure element after the initialization of Google Wallet:

1. A000000003 5350
2. A000000004 10
3. A000000476 10
4. A000000476 1000
5. A000000476 1001
6. A000000476 1002
7. A000000476 20
8. A000000476 30
9. 785041592E

Load-file 1 is assumed to be the load-file used to instantiate security domains (cf. [11]). Load-file 2 contains the MasterCard credit card application and load-file 9 is assumed to contain the EMV payment system environment. Load-file 7 contains the Google Wallet on-card component. Load-file 8 contains Google's application for MIFARE access. Load-files 3–6 have Google AIDs. Their purpose is unknown.

After successful initialization of Google Wallet, several applets can be found on the secure element. The following is a list of applet instance identifiers (AIDs) that have been identified to be used by either Google Wallet during normal operation or by POS terminals during payment transactions:

1. A000000476 2010
2. A000000476 3030
3. 325041592E5359532E4444463031
4. A000000004 1010
5. A000000004 1010 AA54303200FF01FFFF

AID 1 is the Google Wallet on-card component, which is used by Google Wallet to manage the payment cards on the secure element. AID 2 is the Google MIFARE access applet, which is used to manage the MIFARE Classic 4K memory of the secure element. Both applet instances can only be accessed from the application processor. Selecting them though the radio frequency (RF) interface results in the error code 6999. This denotes that the selection of those applets failed.

The last three applet instances are related to credit card payment transactions. AID 3 is the directory definition file of the EMV Proximity Payment System Environment (PPSE) as mandated by [2]. It contains a list of all activated credit card applications. It can be selected from the application processor and through the RF interface. While Google Wallet is in locked state an empty list is returned, otherwise the list contains AID 4 and 5.

AID 4 is the AID for a regular MasterCard credit card. We assume that the other credit card AID (AID 5) denotes a MasterCard co-branded as Google Wallet but we have no confirmation for this.[4] Both credit card applet instances contain equal data structures (including the same primary account number, PAN). The credit card applets are only selectable while Google Wallet is in unlocked state, otherwise the error code 6999 is returned. Access to both applet instances is possible from the application processor and through the RF interface.

7.1.3 Interacting with the Google Wallet On-card Component

Several commands used by the Google Wallet app to interact with its on-card component have been identified:

- 00 A4 0400 07 **A0000000476 2010** 00:
  This SELECT command selects the Google Wallet on-card component by its AID. The secure element responds with 9000 indicating successful selection.
- 80 E2 00**AA** 00:
  This command is used to unlock the wallet and to allow access to the credit card application after the Google Wallet app has successfully verified the PIN. The PIN itself is *not* passed to the on-card component for verification.
- 80 E2 00**55** 00:
  This command is used to lock the wallet.
- 80 CA 00**A5** 00:
  This GET_DATA command returns a list containing the two credit card instances:
  A5 2B
      61 0F (Application template)
        80 01 00
        C2 01 81
        4F 07 (Application identifier)
          A000000004 1010
      61 18 (Application template)
        80 01 01

---

[4] An installation with version 1.5-R79-v5 of the Google Wallet app and version 1.6 of the on-card component installed in September 2012 reports the second credit card AID as A000000004 1010 AA539648FFFF00FFFF.

7.1 Google Wallet   151

      C2 01 81
      4F 10 (Application identifier)
          A000000004 1010 AA54303200FF01FFFF
   9000 (Status: Success)

- 80 F0 **01**00 12 4F 10 **A000000004 1010 AA54303200FF01FFFF** 00: This command is used when the Google Prepaid card is disabled through the Google Wallet app. The secure element returns 61 00 9000.

- 80 F0 **02**00 12 4F 10 **A000000004 1010 AA54303200FF01FFFF** 00: This command is used when the Google Prepaid card is enabled through the Google Wallet app. The secure element returns 61 00 9000.

7.1.4 Google Prepaid Card: A MasterCard PayPass Card

The credit card applet is based on the *EMV Contactless Specifications for Payment Systems* and is a MasterCard PayPass card. It supports Mag-Stripe mode with dynamic CVC3 (card verification code) and requires online transactions. Full EMV mode is not supported. Also, there is no cardholder verification, even though PIN-based cardholder verification would significantly complicate relay attacks or would even render them impossible if the attacker is unable to "guess" the PIN.

A typical Mag-Stripe mode transaction with the Google Prepaid card consists of the following command sequence (cf. [15] for a detailed analysis of a Mag-Stripe transaction):

1. POS → Card: The POS terminal selects (SELECT command) the PPSE:
   00 A4 0400 0E **325041592E5359532E4444463031** 00
2. Card → POS: The secure element responds with the file control information template (FCI) that contains a list of supported EMV payment applications and their priority indicators:
      6F 3A (FCI template)
         84 0E (DF name)
            325041592E5359532E4444463031 ("2PAY.SYS.DDF01")
         A5 28 (Proprietary information encoded in BER-TLV)
            BF0C 25 (FCI issuer discretionary data)
               61 15 (Application template)
                  4F 10 (Application identifier)
                     A000000004 1010 AA54303200FF01FFFF
                  87 01 **01** (Application priority indicator)
               61 0C (Application template)
                  4F 07 (Application identifier)
                     A000000004 1010
                  87 01 **02** (Application priority indicator)
      9000 (Status: Success)

3. POS → Card: The POS terminal selects (SELECT command) the MasterCard Google Prepaid card:
    00 A4 0400 10 **A0000000041010AA54303200FF01FFFF** 00
4. Card → POS: The secure element responds with an FCI that contains application details:
    6F 20 (FCI template)
       84 10 (DF name)
          A0000000041010AA54303200FF01FFFF
       A5 0C (Proprietary information encoded in BER-TLV)
          50 0A (Application label)
             4D617374657243617264 ("MasterCard")
    9000 (Status: Success)
5. POS → Card: The POS terminal retrieves the processing options of the credit card application (GET_PROCESSING_OPTIONS command):
    80 A8 0000 02 8300 00
6. Card → POS: The credit card applet responds with the application interchange profile (**0000** indicates Mag-Stripe mode only, online transactions only, no cardholder verification, etc.) and the location of the Mag-Stripe data file:
    77 0A (Response message template)
       82 02 (Application interchange profile)
          0000
       94 04 (Application file locator)
          08 01 01 00 (short EF = 1, first record = 1, last record = 1)
    9000 (Status: Success)
7. POS → Card: The POS terminal reads (READ_RECORDS command) the Mag-Stripe data from record 1 of the record data file with the short EF 1:
    00 B2 **010C** 00
8. Card → POS: The credit card applet responds with the Mag-Stripe version, track 1 and track 2 information:
    70 6A (Non inter-industry nested data object template)
       9F6C 02 (Mag-Stripe application version number)
          0001 (Version 1)
       9F62 06 (Track 1 bit map for CVC3)
          000000000038
       9F63 06 (Track 1 bit map for UN and ATC)
          0000000003C6
       56 29 (Track 1 data)
          42 (ISO/IEC 7813 structure "B" format)
          35343330 xxxxxxxx 30xxxx37 xxxxxxxx
             (Primary account number "5430 xxxx 0xx7 xxxx")
          5E (Field separator "^")
          202F (Cardholder name "_/")
          5E (Field separator "^")
          31373131 (Expiry date "17"/"11")
          313031 (Service code "101")

7.1 Google Wallet

       303031303030303030303030
       (Discretionary data "0010000000000")
   9F64 01 (Track 1 number of ATC digits)
   04
   9F65 02 (Track 2 bit map for CVC3)
   0038
   9F66 02 (Track 2 bit map for UN and ATC)
   03C6
   9F6B 13 (Track 2 data)
      5430 xxxx 0xx7 xxxx (Primary account number)
      D (Field separator)
      1711 (Expiry date)
      101 (Service code)
      0010000000000 (Discretionary data)
      F (Padding)
   9F67 01 (Track 2 number of ATC digits)
   04
   9000 (Status: Success)

9. POS → Card: The POS terminal instructs the card to compute the cryptographic checksum for a given unpredictable number **nnnnnnnn** (COMPUTE_CRYPTOGRAPHIC_CHECKSUM command):

   80 2A 8E80 04 **nnnnnnnn** 00

10. Card → POS: The credit card applet responds with the application transaction counter (**xxxx**) and with the dynamically generated CVC3 for track 1 (**yyyy**) and track 2 (**zzzz**):

   77 0F (Response message template)
      9F61 02 **zzzz** (CVC3 Track 2)
      9F60 02 **yyyy** (CVC3 Track 1)
      9F36 02 **xxxx** (Application transaction counter, ATC)
   9000 (Status: Success)

Most of the data exchanged in a Mag-Stripe transaction is static for all transactions (e.g. the Mag-Stripe data). COMPUTE_CRYPTOGRAPHIC_CHECKSUM (9 and 10) is the only APDU command-response pair that contains dynamically generated data that differs for each transaction: the unpredictable number generated by the POS and the transaction counter and CVC3 codes generated by the card. Each COMPUTE_CRYPTOGRAPHIC_CHECKSUM command that is sent to the card must be preceded by a fresh GET_PROCESSING_OPTIONS command (5 and 6). Thus, the minimum sequence for generating a dynamic CVC3 is

1. SELECT the MasterCard Google Prepaid card (3 and 4),
2. GET_PROCESSING_OPTIONS (5 and 6), and
3. COMPUTE_CRYPTOGRAPHIC_CHECKSUM (9 and 10).

7.2 Performing a Software-Based Relay Attack

In order to perform a software-based relay attack on a Google Wallet installation, the proof-of-concept implementation used for delay measurements (see Sect. 6.4.3) needed some minor modifications:

The main difference between the delay measurement scenario and access to the credit card in Google Wallet is that Google Wallet is protected by a PIN. However, knowledge of the PIN itself is not necessary to unlock the wallet. Instead, a simple unlock command needs to be sent to the Google Wallet on-card component. After this, the default payment card can be accessed through the internal mode of the secure element. No user interaction is required. If other cards besides the MasterCard Google Prepaid Card are installed into the wallet, additional commands might be necessary to enable the desired payment card.

Thus, whenever the card emulation server begins relaying a transaction, the relay app first selects the Google Wallet on-card component and sends the unlock command. This imitates the behavior of the Google Wallet app upon successful PIN entry by the user. After a transaction completed, the Google Wallet on-card component is selected again and the lock command is used to lock the wallet.

As Android 2.3.7 has been used to initially test the relay scenario with Google Wallet, the access restrictions of the secure element API were loosened in a customized build of the Android firmware. For tests on Android 4.0.3 and 4.1.1, the signature of the relay app was simply added to the secure element permissions file (`/system/etc/nfcee_access.xml`). Root access to the device was necessary in all cases. Instead of manually granting the permissions, privilege escalation exploits could be integrated into future versions of the relay app to automate this process. For easy integration of future exploits, a privilege escalation framework (cf. [10]) could be embedded into the app.

We successfully tested the software-based relay attack with Google Wallet by paying at a real POS terminal with our card emulator (Figs. 7.1 and 7.2). The POS terminal used for these tests was a Hypercom Artema Hybrid with a ViVOtech ViVOpay 5000 contactless reader. For ethical reasons we used our own credit card

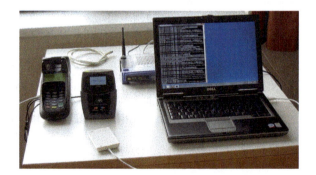

Fig. 7.1 Test setup for performing a payment transaction with the card emulator at a POS terminal (*Source* [15, 17])

7.2 Performing a Software-Based Relay Attack   155

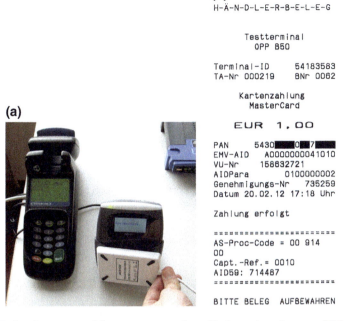

Fig. 7.2 Performing a successful payment transaction with the card emulator at a POS terminal: **a** payment transaction in progress (*Source* [15]), **b** payment receipt

terminal instead of a POS installation in the field. However, the POS terminal is identical to those used in recent roll-outs at Schlecker and Zielpunkt in Austria. Videos of the successful relay attack are available on YouTube [13, 14].

7.3 Viability, Limitations and Improvements

An NFC reader device (available for less than EUR 50), a notebook computer and some programming skills are all that was necessary to mount this attack. However, it has to be admitted that, while using the ACR 122U together with a notebook computer worked in a controlled environment, this setup will certainly raise suspicions when used to pay in a store. An alternative approach would be to use another mobile phone as card emulator. Francis et al. [3] showed that a credit card can be emulated using a BlackBerry 9900 in software card emulation mode (cf. also [16]). The recent addition of host-based card emulation to Android enables software card emulation on a broad range of Android devices. A mobile phone has several advantages:

- accepted form factor for mobile contactless transactions,
- same network interfaces as the device-under-attack, and
- various Android smart phones with NFC (e.g. the Nexus 4) are available for less than EUR 300.

7.3.1 Getting the Relay App on Devices

To roll out the relay app to devices of actual users, it could be integrated into any existing app downloaded from Google Play Store (cf. Sect. 4.5 and Höbarth [9]). The infected app could then be re-published on Google Play Store under similar (or even identical) publisher information and with the same app name as its original. The publisher account that is necessary to re-publish the app costs USD 25 (approximately EUR 20) [5].

For many users it would be difficult to distinguish the original app from the malware, as these apps would only differ in the number of installs and user comments. To specifically target users of rooted devices, an app that already requires root permissions could be used as a base for code injection. This would also simplify root-access, as users would explicitly grant the root privileges to such an app.

However, Google started to combat this approach with updates to the Google Play Developer Program Policy in late 2012.

7.3.2 Transaction Limits

In Austria, PIN-less contactless transactions are usually limited to EUR 25. However, reports on the Internet [12] suggest that Google Wallet can be used for transactions of at least up to USD 100 (approximately EUR 75). An attacker would typically not attack a single Google Wallet device, but instead distribute transactions on many devices infected with the relay app. Thus, an attacker could build a "bot network" of Google Wallets. This method has the advantage that each wallet would be charged less, which might cover the attack for a longer period. Also, the attacker could use the "bot network" to select a device with a good (i.e. stable and fast) network connection.

7.3.3 Optimizing the Relayed Data

The analysis of communication delays induced by a relay attack reveals that the relay channel adds a significant portion of the overall command-response delay. This results in a noticeable slow-down of relayed transactions in comparison to direct transactions. Using the full procedure for a payment transaction described in Sect. 7.1.4, 5 commands (totaling 65 bytes) and 5 responses (totaling 241 bytes) would need to be exchanged between the relay app and the card emulator. Even when only the minimum sequence of commands described in Sect. 7.1.4 is used, this would still result in 3 commands (totaling 40 bytes) and 3 responses (totaling 69 bytes) that need to be exchanged over the relay channel.

One possibility to further improve the speed of relayed transactions is to cache all static transaction data and only transmit dynamically generated data during the

7.3 Viability, Limitations and Improvements

transaction. Thus, it is sufficient to transmit the dynamic fields of the COMPUTE_CRYPTOGRAPHIC_CHECKSUM command-response pair over the relay channel. All other data can be retrieved from the Google Wallet device prior to the attack. This reduces the number of commands exchanged over the relay channel during one transaction to one command (containing a 4 byte unpredictable number as the data) and to one response (containing 6 bytes of data generated by the secure element in response to the unpredictable number). As a consequence, even a single SMS message for each direction would be sufficient as a relay channel if all other information (e.g. Mag-Stripe data) has been previously cached on the card emulator.

7.4 Possible Workarounds

Section 6.5 gives an overview of possible solutions to prevent relay attacks and software-based relay attacks in particular. Several of these solutions could be applied to Google Wallet. However, each method has its advantages and disadvantages.

7.4.1 Timeouts of POS Terminals

An easy, but potentially unreliable, measure to prevent relay attacks would be the enforcement of short timeouts (e.g. the benchmark targets specified by the EMV specifications) for payment transactions on the POS terminals. Transactions taking longer than this timeout should be interrupted or discarded. While this measure will prevent most long-distance relay scenarios, relays over shorter distances and fast communication channels might not be rejected. Also, such tight timeouts will prevent some cloud-based EMV applications (cf. Sect. 6.5).

7.4.2 Google Wallet PIN Verification

> A PIN and the ability to remotely disable Google Wallet make it very safe [8].

In version 1.1-R52v7 of the Google Wallet app, the PIN that protects the wallet is only verified within the mobile phone app. Simple lock and unlock commands are used to control the state of the on-card component instead of on-card PIN verification. This, once more, delegates access control for a secure component (Google Wallet on-card component and credit card applets) to a potentially insecure component (Google Wallet app on the application processor). The on-card component does not verify this PIN.

PIN verification could be handled by the on-card component on the secure element. After all, PIN verification is a core component of smartcards anyways. In that case,

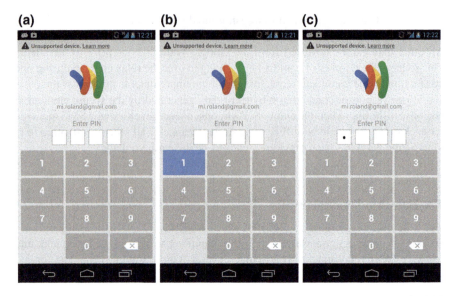

Fig. 7.3 Monitoring the Google Wallet PIN entry dialog may allow an attacker to capture the PIN: **a** PIN entry dialog, **b** touch event for digit "1" causes the button to change its color from *gray* to *blue*, **c** digit "1" has been released

the attacker would need to know the wallet PIN in order to conduct a successful attack. Still, a malicious app with sufficient privileges might be able to monitor PIN entry (e.g. by intercepting keyboard inputs or by capturing the screen on touch events, cf. Fig. 7.3).

Another approach would be to require PIN entry and online PIN verification at the point-of-sale for any transaction amount. However, this is impracticable or even impossible at certain points-of-sale.

7.4.3 Disabling Internal Mode for Payment Applets

Modern secure elements (like those embedded into Google's Nexus devices) provide instruments to distinguish between external mode communication and internal communication from within a Java Card applet. Rules for interface-based access can be applied on a per-applet basis and even on a per-APDU basis. These capabilities could be used to disable internal mode communication for all payment applets and consequently disable their vulnerability for software-based relay attacks.

The disadvantage of this workaround is that the secure element cannot be used for future on-device secure payment applications (e.g. EMV-based authorization of payment transactions in the mobile phone web browser). Such applications would, however, be one of the key benefits of having a secure element inside a mobile phone.

7.5 Reporting and Industry Response

We reported our findings and proposed workarounds to Google (and some of their Google Wallet partners) in April 2012. Google quickly acknowledged the problem and confirmed that they could reproduce the attack. Our tests in June 2012 revealed that new installations of Google Wallet (i.e. secure element applets provisioned in June) were no longer vulnerable to our relay attack setup. Further testing in September 2012 showed that users of older versions of Google Wallet are now required to update to the latest version. This forces existing users to receive the necessary fixes of the secure element applets. Therefore, we assume that Google Wallet users are no longer vulnerable to the relay attack scenario described in this thesis.

Google acknowledged the report of this security vulnerability with an entry in the "Honorable Mention" section of their *Application Security Hall of Fame* [6].

7.6 Analysis of the Relay-Immune Google Wallet

Version 1.6 of the Google Wallet on-card component (installed with version 1.5-R79-v5 of Google Wallet in September 2012, cf. Fig. 7.4) is no longer vulnerable to the software-based relay attack setup described in this thesis. The relay attack is inhibited by the fact that the select commands for both MasterCard credit card applet

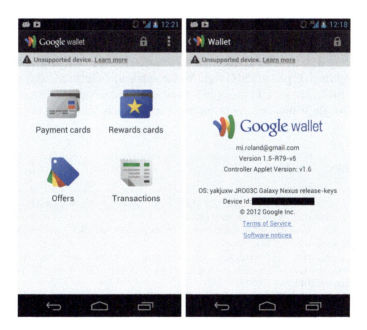

Fig. 7.4 Google Wallet version 1.5-R79-v5

instances[5] fail with the error code 6999. Thus, access to the credit card applet from the application processor has been disabled as we suggested (cf. Sect. 7.4.3). The PPSE can still be selected through both internal and external mode.

The on-card component can still only be selected through internal mode. It now returns its version number upon selection:

6F 0F (FCI template)
    84 07 (DF name)
        A000000476 2010
    A5 04 (Proprietary information encoded in BER-TLV)
        80 02 **0106** (Version 1.6)
9000 (Status: Success)

Also, some commands for interaction with the on-card component have slightly changed their parameters (e.g. the commands for enabling and disabling a specific payment card). However, the commands for switching between locked and unlocked state of the wallet are still the same. As a result, it is still possible to unlock Google Wallet and the credit card contained in it without PIN verification. Consequently, a malicious application with access to the secure element could enable the credit card on the RF contactless interface even though Google Wallet is protected by a PIN that is not known to the malicious application.

In a thread on XDA Developers [18], Rubin explains why PIN verification has not been moved to the secure element:

> Google believes that the change required may constitute a 'change of agency' regarding who does the PIN verification (if it is done inside the secure element). If the banks then become responsible for the PIN verification, the PIN becomes subject to the same regulations and procedures as an ATM PIN.

However, in my opinion, this would only apply if the PIN was part of the credit card application. If the PIN verification is performed in the Google Wallet on-card component, which is part of the Google Wallet application, responsibility would not shift to the banks but would remain with Google and the Google Wallet application. After all, Google Wallet only *hosts* the credit card applets for the banks.

References

1. Drake, J.J., Oliva Fora, P., Lanier, Z., Mulliner, C., Ridley, S.A., Wicherski, G.: Android Hacker's Handbook. Wiley, New York (2014)
2. EMVCo: EMV Contactless Specifications for Payment Systems—Book B: Entry Point Specification. Version 2.1 (2011)
3. Francis, L., Hancke, G.P., Mayes, K.E., Markantonakis, K.: Practical relay attack on contactless transactions by using NFC mobile phones. Cryptology ePrint Archive, Report 2011/618. http://eprint.iacr.org/2011/618 (2011)
4. GlobalPlatform: Card Specification. Version 2.2.1 (2011)

---

[5] AIDs A000000004 1010 and A000000004 1010 AA539648FFFF00FFFF.

References

5. Google: Android developer—Google Play developer help—developer registration. https://support.google.com/googleplay/android-developer/answer/113468 (2014). Accessed Dec 2014
6. Google: Google—application security—hall of fame—honorable mention. http://www.google.com/about/appsecurity/hall-of-fame/distinction/ (2014). Accessed Dec 2014
7. Google: Google Wallet. https://www.google.com/wallet/ (2012). Accessed Sept 2012
8. Google: Google Wallet—how it works—in-store. http://www.google.com/wallet/how-it-works/in-store.html (2012). Accessed Sept 2012
9. Höbarth, S.: Android monkeys—get it, malware it, market it. Presentation at Hacking Night WS 2011. Hagenberg, Austria (2012)
10. Höbarth, S., Mayrhofer, R.: A framework for on-device privilege escalation exploit execution on Android. In: 3rd International Workshop on Security and Privacy in Spontaneous Interaction and Mobile Phone Use. San Francisco, CA, USA. http://www.medien.ifi.lmu.de/iwssi2011/papers/hoebarth-spmu2011.pdf (2011)
11. Mostowski, W., Pan, J., Akkiraju, S., de Vink, E., Poll, E., den Hartog, J.: A comparison of Java Cards: state-of-affairs 2006. CS-Report CSR 07–06, Technische Universiteit Eindhoven (2007)
12. Planck, S.: Google Wallet statistics roundup. NFC rumors. http://www.nfcrumors.com/05-27-2011/google-wallet-statistics-roundup/ (2011)
13. Roland, M.: Google Wallet relay attack. http://youtu.be/_R2JVPJzufg
14. Roland, M.: Google Wallet relay attack (low quality). http://youtu.be/hx5nbkDy6tc
15. Roland, M.: Applying recent secure element relay attack scenarios to the real world: Google Wallet relay attack. Comput. Res. Repository (CoRR), arXiv:1209.0875 (cs.CR) (2012). http://arxiv.org/abs/1209.0875
16. Roland, M.: Software card emulation in NFC-enabled mobile phones: great advantage or security nightmare? In: 4th International Workshop on Security and Privacy in Spontaneous Interaction and Mobile Phone Use. Newcastle, UK. http://www.medien.ifi.lmu.de/iwssi2012/papers/iwssi-spmu2012-roland.pdf (2012)
17. Roland, M., Langer, J., Scharinger, J.: Applying relay attacks to Google Wallet. In: Proceedings of the Fifth International Workshop on Near Field Communication (NFC 2013). IEEE, Zurich, Switzerland (2013). doi:10.1109/NFC.2013.6482441
18. Rubin, J.: Google wallet PIN vulnerability, post #5 on 9 Feb 2012 12:45 AM by J. Rubin (alias "miasma"). Thread on XDA Developers forum. http://forum.xda-developers.com/showpost.php?p=22327658&postcount=5 (2012). Accessed Sept 2012
19. Rubin, J.: Google Wallet security: about that rooted device requirement... zveloBLOG. https://zvelo.com/blog/entry/google-wallet-security-about-that-rooted-device-requirement (2012)

Chapter 8
Summary and Outlook

This work assessed the current state of Near Field Communication (NFC) security with regard to a range of specific application scenarios. Based on exemplary use-cases from the area of improving efficiency in automotive environments, application-specific security requirements have been identified. Two aspects of NFC—tagging and card emulation—have been found to be particularly important. Both aspects have been evaluated with regard to the efficiency of existing security architectures. Weaknesses of the existing security measures and new attack scenarios have been identified for both, tagging and secure element based card emulation. Countermeasures and solutions to overcome these unresolved security issues have been outlined.

8.1 Tagging

For the interaction with NFC tags (i.e. the tagging scenario), the NFC Forum published the NDEF Signature Record Type Definition specification as a first approach towards achieving authenticity and integrity for data stored on NFC tags and for data transferred across the NFC link. Nevertheless, this specification is only a small step towards authenticity and integrity protection. The signature record type definition only specifies a container record format for a digital signature and rules for generating that digital signature. Another important part—the infrastructure behind the digital signatures and the certificates—has yet to be defined. A public-key infrastructure and rules for establishing trust in signature issuers and signed data are an essential part of the overall signature system. Therefore, a possible design of a public key infrastructure (PKI) for digital signature of NDEF data has been outlined. Moreover, the advantages and disadvantages of different methods for binding digital certificates to signed content have been discussed.

Two critical design criteria that have to be considered carefully when implementing an NDEF-PKI are the management of the private signing keys and the lifespan of certificates. Several different models for both design criteria have been evaluated with regard to their advantages and disadvantages.

A promising approach for the management of private signing keys seems to be online signature generation at an online signature generation service owned by the certificate authority (CA). While that approach requires a considerable amount of trust in that centralized service, it significantly reduces complexity for the content issuer. The content issuer does not need to handle the secret key which results in reduced costs. At the same time, the content issuer can even delegate signature generation to a tag manufacturer without passing on the secret key.

With regard to the lifespan and the validity period of certificates and signatures, several existing models have been evaluated for their viability in the tagging scenario. The evaluation revealed that there is a notable trade-off between long-term usability of signed NFC tags and the security of their signatures. In particular the expected lifetime of NFC tags varies between different applications. In some cases the expected lifespan of NFC tags is still not clear as the mass-adoption of NFC is only slowly beginning.

Though the signature RTD only specifies the container for transporting digital signatures and the rules for computing a signature, an in-depth analysis revealed that the specification even fails its goal of providing an adequate means for protecting the authenticity and integrity of NDEF records. The signature RTD contains a vulnerability that permits several methods of manipulating the content of signed NDEF records without voiding their signature. The weakness is that only parts of the fields of signed records are covered by the signature while important header fields are not protected by the signature. Consequently, the signature RTD fails to provide adequate integrity protection.

The vulnerabilities can be abused to conduct a "record composition attack". The record composition attack is an attack scenario where multiple existing signed NDEF messages are aggregated into a new NDEF message that conveys a new meaning. Unwanted parts of the original messages are selectively hidden from the new message by manipulating header fields that are not covered by the signature.

Besides its vulnerability to hiding and manipulation of signed NDEF records, the signature RTD is also susceptible to security and privacy issues caused by the use of remote URIs (uniform resource identifiers) as part of the signature records. These URIs have no form of authenticity and integrity protection. Therefore, they could be freely manipulated by an attacker. Possible attack scenarios range from collecting usage information for NFC tag infrastructures to triggering HTTP GET requests for arbitrary URIs in the context of the user.

8.2 Card Emulation

Secure element based card emulation is often considered secure due to its use of secure smartcard technology. Also, many secure element access APIs have sophisticated access control mechanisms to prevent unauthorized applications from accessing the secure element. However, an analysis performed as part of this thesis showed that all APIs rely on the integrity of the operating system when performing access control decisions. Once an application is able to elevate its privileges (e.g. to gain root privileges), it is also capable of circumventing the access control policies of secure element APIs.

While applications on the secure element are usually protected with further security measures such as shared keys and encryption, some attack scenarios are possible even if the data stored inside the secure element applications itself is inaccessible. In particular, two potential attack scenarios have been identified: a denial-of-service attack and a software-based relay attack.

The denial-of-service attack uses the security mechanisms of the GlobalPlatform card management interface to either permanently block any further card management operations (e.g. installation and removal of applets) or to temporarily block all further communication with the secure element.

The software-based relay attack is an extension of existing relay attack concepts. Traditional relay attacks on contactless smartcards and secure element enabled mobile phones rely on physical proximity between the device-under-attack and the attacker. Also, external card emulation needs to be enabled during the attack. However, the software-based relay attack no longer has these restrictions. Especially physical proximity to the device-under-attack is not required. Instead, an app on the mobile phone application processor relays the communication between an attacker and the secure element over the cellular network or another wireless communication channel. That way, an attacker could build a card emulator (*proxy*) that relays the commands received from a real reader device (e.g. a point-of-sale (POS) terminal) through the relay app on the device-under-attack to the secure element and that routes the responses of the secure element back to the reader.

Protocol analysis and reference measurements with a prototype implementation of the relay system revealed that such relay attacks are possible and not hindered by any of the protocols involved in the communication. Specifically, the fact that the communication with the secure element is performed on the application layer (APDUs, application protocol data units) further relaxes the timing constraints.

To verify the software-based relay attack with an existing application, Google Wallet has been chosen as an attack target. An analysis of the communication of the Google Wallet app with its on-card component in the secure element and an analysis of the communication between Google Wallet and an actual POS terminal during a payment transaction showed that the MasterCard-branded prepaid credit card stored inside the wallet is based on EMV payment card standards. The analysis also revealed that the credit card is protected inside the wallet by activation and deactivation commands. The activation command is necessary before the payment card can be accessed and is usually sent by the Google Wallet app upon successful PIN entry.

After modifying the prototype relay system to issue the activation command before starting the relay communication, the relay attack could be successfully used to perform a payment transaction at a real POS terminal with the Google MasterCard prepaid card stored inside the wallet. These findings, together with a number of proposals for possible workarounds, have been reported to Google and some of their wallet partners. Google responded by releasing an updated version of their wallet that rejects access to the credit card application on the secure element from apps on the mobile phone application processor. This fix matches one of our proposed solutions. Google acknowledged the responsible disclosure of this vulnerability of their wallet with an entry in their *Application Security Hall of Fame* [3]. The story of the vulnerability has even been picked up by the media (e.g. derStandard [12], Die Presse [2], Futurezone [13], Oberösterreichische Nachrichten [4], ORF.at [11]).

8.3 Conclusion

NFC technology has some security measures for all three of its operating modes. For reader/writer mode and tagging, write protection of tags has been the only security measure for several years. The recently released signature RTD adds digital signature as an additional security measure. However, due to the lack of a PKI, most NDEF applications do not use digital signatures yet. For peer-to-peer mode there is an ISO standard that provides a secure communication channel for NFC. Nonetheless, this secure channel protocol is not used in any of the current smart phone peer-to-peer stacks. There are also no NFC Forum specifications based on this standard. For card emulation mode there is the secure element as central security measure. As the secure element is based on secure smartcard technology, it has similar security properties as regular smartcards. Nevertheless, the secure element has insufficient protection against unauthorized access on most current mobile phone platforms.

8.4 The Bigger Picture

While NFC has been developed in 2002 and the first NDEF specifications appeared in 2006, it took four more years until the NFC Forum released its first security specification: the Signature Record Type Definition. Even today, 4 years after the signature RTD was initially published, it is still the NFC Forum's only security related specification. Only in late 2014 a policy document [10] with guidelines for a signature RTD certification authority has been released. As a result, digital signatures for NDEF are still not widely used.

Nevertheless, more and more mobile phones are equipped with NFC and more and more NFC tags are in use. Also, several installations of NFC tag infrastructures are already in place. Therefore, upgrading to signature-based security would require all these existing tags to be replaced with tags that contain valid signatures. Otherwise, these tag infrastructures would have a significantly decreased user-experience (e.g. additional warning messages and additional user interaction). The longer it takes to publish a complete security specification and guidelines to NFC security, the more products and services without security features will become available. As a consequence, it will become difficult to convince service providers that additional security measures are even necessary, especially as the cost for these measures increases with the size of their existing tag infrastructures.

A similar situation exists with NFC devices. While there have been various publications about vulnerabilities in the first NFC-enabled mobile phones (e.g. Mulliner [7, 8] in 2008/2009), similar security issues still exist in current Android NFC devices (cf. Miller [6], Mulliner [9], Benninger and Sobell [1]). Thus, years after the first reports about vulnerabilities in NFC devices, these very same vulnerabilities are still built into new devices. Therefore, even 10 years after the birth of the NFC technology, there still seem to be no guidelines that device manufacturers can follow to build devices that are robust to well-known attacks.

For the card emulation scenario, the first platform independent API specification for secure element access has been released in 2011: the Open Mobile API. It took one more year until a standard for access control to the secure element was published. Before, only standards for Java ME (Java Platform, Micro Edition) and several proprietary interfaces existed. Nevertheless, even these new specifications make the assumption that the mobile phone platform (hardware, operating system, etc.) itself is secure. However, many publications (cf. Sect. 4.5) prove that this cannot be assumed for most of the current smart phones.

In all these cases, security related specifications, standards and guidelines have been developed in a late stage only after the technology itself became more popular or have not been developed at all. This seems to be a significant flaw in the strategies around NFC technology. It seems as if there is a priority on pushing standards for interoperable applications while security is left aside as a low priority. Development of security standards seems to happen only in response to actual vulnerability reports and threats. However, NFC security could potentially be implemented in a continuous proactive process that happens in parallel to the development of the protocol and application specifications. This would particularly help service providers and application developers that create applications and services based on these standards to implement security from the beginning.

8.5 Future Research

This thesis identified several issues in current security measures for NFC applications and gave an overview of possible countermeasures. Nevertheless, with both, tagging and card emulation, there are still several questions left for future research.

Particularly the various options for a public-key infrastructure for digital signature of NDEF messages need further evaluation. For instance, certificate binding needs to be analyzed based on existing NDEF applications and for combinations of multiple records. Another topic for future research is user interaction and the implementation of signature verification in general and on specific device platforms. Korak and Wilfinger [5] built an Android implementation of the signature RTD that can detect potentially malicious signed NDEF records based on the weaknesses identified in this thesis. Their implementation also provides a first approach to differentiated user interaction based on signature verification results.

For the integration of NFC card emulation and secure elements into smart phones, the focus of future research could be on secure authentication of apps to the secure element. For instance, trusted platform concepts seem to be a promising approach to reliably perform access control to the secure element. Trusted platform concepts could also provide a basis for trusted user interaction with the secure element (e.g. to provide a trusted display and trusted PIN entry that cannot be intercepted by malicious applications). Trusted input/output capabilities could also be used to feed the secure element with trusted sensor data (e.g. GPS) that could be used to implement countermeasures against relay attacks comparable to distance bounding and packet leashes.

Nevertheless, future research might also search for new attack surfaces. For example, other applications on the secure element besides the card manager might also be vulnerable to denial-of-service attacks. In particular applications that use PIN codes could be vulnerable to intentionally exceeding the limits of the retry counters. Also EMV-compliant credit cards contain transaction counters to prevent replay attacks. These counters could possibly be incremented until they exceed their maximum value which might make a payment card unusable for further transactions.

Further more, a recent trend towards host-based card emulation (HCE) in combination with cloud-based secure elements (as an alternative to on-device secure elements) could potentially open new attack surfaces. Both, the cloud-based secure element and the HCE app on the mobile device provide interfaces similar to those of a secure element. Consequently, it may be possible to extend the software-based relay attack scenario to HCE applications. An attacker could, for instance, try to hijack the communication channel between the wallet app and the cloud-based secure element. Alternatively, an attacker could try to inject communication into the HCE interface of an app.

References

1. Benninger, C., Sobell, M.: Intro to Near Field Communication (NFC) mobile security. Presentation at ShmooCon 2012. Washington, DC, USA. http://youtu.be/ZWQKV0DI2jw (2012)
2. Die Presse: Linzer Forscher löst Sicherheitsproblem für Google. DiePresse.com. http://diepresse.com/home/techscience/mobil/android/1304511/ (2012)
3. Google: Google—Application Security—Hall of Fame—Honorable Mention. http://www.google.com/about/appsecurity/hall-of-fame/distinction/ (2014). Accessed Dec 2014
4. Habringer, A.: Drei Buchstaben beherrschen seine Welt. Oberösterreichische Nachrichten. http://www.nachrichten.at/oberoesterreich/art4,996318 (2012)
5. Korak, T., Wilfinger, L.: Handling the NDEF signature record type in a secure manner. In: Proceedings of the IEEE International Conference on RFID-Technologies and Applications (RFID-TA 2012), pp. 107–112. IEEE, Nice, France (2012). doi:10.1109/RFID-TA.2012.6404492
6. Miller, C.: Don't Stand So Close To Me: An Analysis of the NFC Attack Surface. Briefing at BlackHat USA. Las Vegas, NV, USA (2012)
7. Mulliner, C.: Attacking NFC Mobile Phones. Talk at 25th Chaos Communication Congress. Berlin, Germany. http://www.mulliner.org/nfc/feed/collin_mulliner_25c3_attacking_nfc_phones.pdf (2008)

8. Mulliner, C.: Vulnerability analysis and attacks on NFC-enabled mobile phones. In: Proceedings of the International Conference on Availability, Reliability and Security (ARES '09), pp. 695–700. IEEE, Fukuoka, Japan (2009). doi:10.1109/ARES.2009.46
9. Mulliner, C.: Binary Instrumentation on Android. Talk at SummerCon. New York, NY, USA. http://www.mulliner.org/android/feed/binaryinstrumentationandroid_mulliner_summercon12.pdf (2012)
10. NFC Forum: Signature RTD Certificate Policy. Policy document, version 1.0 (2014)
11. ORF: Sicherheitslücke beim Bezahlen per Handy. ORF.at. http://ooe.orf.at/news/stories/2555729/ (2012)
12. Pumhösel, A.: Googles Geldtasche gehackt. derStandard.at. http://derstandard.at/1350260526386/Googles-Geldtasche-gehackt (2012)
13. Wimmer, B.: Österreicher deckt NFC-Lücke bei Google auf. futurezone.at Technology News. http://futurezone.at/science/oesterreicher-deckt-nfc-luecke-bei-google-auf/24.586.384 (2012)

Appendix A
Google's Secure Element API

This appendix lists the interface definition of Google's proprietary secure element API as included in version 4.2.1 of the Android system. See Sect. 6.2.4.1 for a detailed analysis.

A.1 Class NfcAdapterExtras

```
1   /*
2    * Copyright (C) 2011 The Android Open Source Project
3    *
4    * Licensed under the Apache License, Version 2.0 (the
5    * "License"); you may not use this file except in compliance
6    * with the License. You may obtain a copy of the License at
7    *
8    *      http://www.apache.org/licenses/LICENSE-2.0
9    *
10   * Unless required by applicable law or agreed to in writing,
11   * software distributed under the License is distributed on
12   * an "AS IS" BASIS, WITHOUT WARRANTIES OR CONDITIONS OF ANY
13   * KIND, either express or implied. See the License for the
14   * specific language governing permissions and limitations
15   * under the License.
16   */
17
18  package com.android.nfc_extras;
19
20  public final class NfcAdapterExtras {
21      /**
22       * Broadcast Action: RF field ON has been detected.
23       * This is an unreliable signal, and will be removed.
24       */
25      public static final String ACTION_RF_FIELD_ON_DETECTED =
26              "com.android.nfc_extras.action.RF_FIELD_ON_DETECTED";
27
28      /**
29       * Broadcast Action: RF field OFF has been detected.
30       * This is an unreliable signal, and will be removed.
31       */
32      public static final String ACTION_RF_FIELD_OFF_DETECTED =
33              "com.android.nfc_extras.action.RF_FIELD_OFF_DETECTED";
```

```java
    /**
     * Get the NfcAdapterExtras for the given NfcAdapter.
     */
    public static NfcAdapterExtras get(NfcAdapter adapter) { ... }


    /**
     * Immutable data class that describes a card emulation route.
     */
    public final static class CardEmulationRoute {
        /**
         * Card Emulation is turned off on this NfcAdapter.
         */
        public static final int ROUTE_OFF = 1;

        /**
         * Card Emulation is routed to nfcEe only when the screen
         * is on, otherwise it is turned off.
         */
        public static final int ROUTE_ON_WHEN_SCREEN_ON = 2;

        /**
         * A route such as ROUTE_OFF or ROUTE_ON_WHEN_SCREEN_ON.
         */
        public final int route;

        /**
         * The NFCExecutionEnvironment that Card Emulation is
         * routed to.
         */
        public final NfcExecutionEnvironment nfcEe;


        public CardEmulationRoute(
                int route, NfcExecutionEnvironment nfcEe) { ... }
    }


    /**
     * Get the current routing state of the secure element.
     */
    public CardEmulationRoute getCardEmulationRoute() { ... }

    /**
     * Set the routing state of the secure element.
     */
    public void setCardEmulationRoute(
            CardEmulationRoute route) { ... }


    /**
     * Get the NfcExecutionEnvironment for the embedded secure
     * element.
     */
    public NfcExecutionEnvironment getEmbeddedExecutionEnvironment(
            ) { ... }


    /**
     * Authenticate the client application.
     * Some implementations of NFC Adapter Extras may require
     * applications to authenticate with a token, before using
     * other methods.
     * This method is not used on Nexus S/Galaxy Nexus.
     */
```

Appendix A: Google's Secure Element API

```
101     public void authenticate(byte[] token) { ... }
102
103     /**
104      * Returns the name of this adapter's driver.
105      */
106     public String getDriverName() { ... }
107 }
```

A.2 Class NfcExecutionEnvironment

```
1   /*
2    * Copyright (C) 2011 The Android Open Source Project
3    *
4    * Licensed under the Apache License, Version 2.0 (the
5    * "License"); you may not use this file except in compliance
6    * with the License. You may obtain a copy of the License at
7    *
8    *      http://www.apache.org/licenses/LICENSE-2.0
9    *
10   * Unless required by applicable law or agreed to in writing,
11   * software distributed under the License is distributed on
12   * an "AS IS" BASIS, WITHOUT WARRANTIES OR CONDITIONS OF ANY
13   * KIND, either express or implied. See the License for the
14   * specific language governing permissions and limitations
15   * under the License.
16   */
17
18  package com.android.nfc_extras;
19
20  import java.io.IOException;
21
22  public class NfcExecutionEnvironment {
23      /**
24       * Broadcast Action: An ISO-DEP AID was selected.
25       */
26      public static final String ACTION_AID_SELECTED =
27          "com.android.nfc_extras.action.AID_SELECTED";
28      /**
29       * Mandatory byte array extra field in ACTION_AID_SELECTED.
30       */
31      public static final String EXTRA_AID =
32          "com.android.nfc_extras.extra.AID";
33
34      /**
35       * Broadcast action: A filtered APDU was received.
36       */
37      public static final String ACTION_APDU_RECEIVED =
38          "com.android.nfc_extras.action.APDU_RECEIVED";
39      /**
40       * Mandatory byte array extra field in ACTION_APDU_RECEIVED.
41       */
42      public static final String EXTRA_APDU_BYTES =
43          "com.android.nfc_extras.extra.APDU_BYTES";
44
45      /**
46       * Broadcast action: An EMV card removal event was detected.
47       */
48      public static final String ACTION_EMV_CARD_REMOVAL =
49          "com.android.nfc_extras.action.EMV_CARD_REMOVAL";
50
```

```java
 51    /**
 52     * Broadcast action: An adapter implementing MIFARE Classic
 53     * via card emulation detected that a block has been accessed.
 54     */
 55    public static final String ACTION_MIFARE_ACCESS_DETECTED =
 56            "com.android.nfc_extras.action.MIFARE_ACCESS_DETECTED";
 57    /**
 58     * Optional integer extra field in ACTION_MIFARE_ACCESS
 59     * _DETECTED that provides the block number being accessed.
 60     */
 61    public static final String EXTRA_MIFARE_BLOCK =
 62            "com.android.nfc_extras.extra.MIFARE_BLOCK";
 63
 64
 65    /**
 66     * Open the NFC Execution Environment on its contact
 67     * interface.
 68     */
 69    public void open() throws IOException { ... }
 70
 71    /**
 72     * Close the NFC Execution Environment on its contact
 73     * interface.
 74     */
 75    public void close() throws IOException { ... }
 76
 77    /**
 78     * Send raw commands to the NFC Execution Environment
 79     * and receive the response.
 80     */
 81    public byte[] transceive(byte[] in) throws IOException { ... }
 82 }
```

Appendix B
Modifications to Google's Secure Element API Library

This appendix lists a modified version of Google's proprietary secure element API that outputs debug information to the Android debug log. This version is based on `com.android.nfc_extras` as included in Android 4.1.1. Most comments were removed from these listings.

B.1 Class NfcAdapterExtras

```
1   /*
2    * Copyright (C) 2011 The Android Open Source Project
3    * Modifications (debug output): (C) 2012 Michael Roland
4    *
5    * Licensed under the Apache License, Version 2.0 (the
6    * "License"); you may not use this file except in compliance
7    * with the License. You may obtain a copy of the License at
8    *
9    *      http://www.apache.org/licenses/LICENSE-2.0
10   *
11   * Unless required by applicable law or agreed to in writing,
12   * software distributed under the License is distributed on
13   * an "AS IS" BASIS, WITHOUT WARRANTIES OR CONDITIONS OF ANY
14   * KIND, either express or implied. See the License for the
15   * specific language governing permissions and limitations
16   * under the License.
17   */
18
19  package com.android.nfc_extras;
20
21  import java.util.HashMap;
22
23  import android.content.Context;
24  import android.nfc.INfcAdapterExtras;
25  import android.nfc.NfcAdapter;
26  import android.os.RemoteException;
27  import android.util.Log;
28  import java.lang.reflect.Method;
```

```
37  public final class NfcAdapterExtras {
38    private static final String TAG = "NfcAdapterExtras";



48    public static final String ACTION_RF_FIELD_ON_DETECTED =



59    public static final String ACTION_RF_FIELD_OFF_DETECTED =
60        "com.android.nfc_extras.action.RF_FIELD_OFF_DETECTED";



65    private static INfcAdapterExtras sService;
66    private static final CardEmulationRoute ROUTE_OFF =
67        new CardEmulationRoute(CardEmulationRoute.ROUTE_OFF,
68                               null);



71    private static final HashMap<NfcAdapter, NfcAdapterExtras>
72        sNfcExtras = new HashMap();
73
74    private final NfcExecutionEnvironment mEmbeddedEe;
75    private final CardEmulationRoute mRouteOnWhenScreenOn;
76
77    private final NfcAdapter mAdapter;
78    final String mPackageName;
79
80    /** get service handles */
81    private static void initService(NfcAdapter adapter) {
82      try {
83        Method getNfcAdapterExtrasInterface =
84            NfcAdapter.class.getMethod(
85                "getNfcAdapterExtrasInterface");
86        final INfcAdapterExtras service = (INfcAdapterExtras)
87            getNfcAdapterExtrasInterface.invoke(adapter);
88        if (service != null) {
89          // Leave stale rather than receive a null value.
90          sService = service;
91        }
92      } catch (Exception e) {}
93    }


104   public static NfcAdapterExtras get(NfcAdapter adapter) {
105     Context context = null;
106     try {
107       Method getContext =
108           NfcAdapter.class.getMethod("getContext");
109       context = (Context)getContext.invoke(adapter);
110     } catch (Exception e) {}
111     if (context == null) {
112       throw new UnsupportedOperationException(
113           "You must pass a context to your NfcAdapter to use the
                NFC extras APIs");
114     }
115
116     synchronized (NfcAdapterExtras.class) {
117       if (sService == null) {
118         initService(adapter);
119       }
120       NfcAdapterExtras extras = sNfcExtras.get(adapter);
121       if (extras == null) {
122         extras = new NfcAdapterExtras(adapter);
```

Appendix B: Modifications to Google's Secure Element API Library

```
123              sNfcExtras.put(adapter,   extras);
124          }
125          return extras;
126       }
127    }
128
129    private NfcAdapterExtras(NfcAdapter adapter) {
130       mAdapter = adapter;
131       String packageName = null;
132       try {
133          Method getContext =
134                NfcAdapter.class.getMethod("getContext");
135          packageName = ((Context)getContext.invoke(adapter))
136                .getPackageName();
137       } catch (Exception e) {}
138       mPackageName = packageName;
139       mEmbeddedEe = new NfcExecutionEnvironment(this);
140       mRouteOnWhenScreenOn = new CardEmulationRoute(
141             CardEmulationRoute.ROUTE_ON_WHEN_SCREEN_ON,
142             mEmbeddedEe);
143    }


148    public final static class CardEmulationRoute {


153       public static final int ROUTE_OFF = 1;


159       public static final int ROUTE_ON_WHEN_SCREEN_ON = 2;


164       public final int route;


170       public final NfcExecutionEnvironment nfcEe;
171
172       public CardEmulationRoute(int route,
173                                 NfcExecutionEnvironment nfcEe) {
174          if (route == ROUTE_OFF && nfcEe != null) {
175             throw new IllegalArgumentException(
176                   "must not specifiy a NFC-EE with ROUTE_OFF");
177          } else if (route != ROUTE_OFF && nfcEe == null) {
178             throw new IllegalArgumentException(
179                   "must specifiy a NFC-EE for this route");
180          }
181          this.route = route;
182          this.nfcEe = nfcEe;
183       }
184    }


189    void attemptDeadServiceRecovery(Exception e) {
190       Log.e(TAG,
191             "NFC Adapter Extras dead - attempting to recover");
192       try {
193          Method attemptDeadServiceRecovery =
194                NfcAdapter.class.getMethod(
195                      "attemptDeadServiceRecovery",
196                      Exception.class);
197          attemptDeadServiceRecovery.invoke(mAdapter, e);
198       } catch (Exception ee) {}
```

```
199       initService(mAdapter);
200     }
201
202     INfcAdapterExtras getService() {
203       return sService;
204     }


212     public CardEmulationRoute getCardEmulationRoute() {
213       try {
214         int route = sService.getCardEmulationRoute(mPackageName);
215         Log.d(TAG,
216             "getCardEmulationRoute() for " +
217             ((mPackageName != null) ? mPackageName : "[null]") +
218             " = " + route);
219         return route == CardEmulationRoute.ROUTE_OFF ?
220             ROUTE_OFF :
221             mRouteOnWhenScreenOn;
222       } catch (Exception e) {
223         attemptDeadServiceRecovery(e);
224         return ROUTE_OFF;
225       }
226     }


238     public void setCardEmulationRoute(CardEmulationRoute route) {
239       try {
240         Log.d(TAG,
241             "setCardEmulationRoute(" + route.route + ") for " +
242             ((mPackageName != null) ? mPackageName : "[null]"));
243         sService.setCardEmulationRoute(mPackageName, route.route);
244       } catch (Exception e) {
245         attemptDeadServiceRecovery(e);
246       }
247     }


258     public NfcExecutionEnvironment getEmbeddedExecutionEnvironment() {
259       Log.d(TAG,
260           "getEmbeddedExecutionEnvironment() for " +
261           ((mPackageName != null) ? mPackageName : "[null]"));
262       return mEmbeddedEe;
263     }
264
265     /**
266      * Convert a byte array into its hexadecimal string form.
267      * @param b   Byte array.
268      * @return    Hexadecimal string representation.
269      */
270     private static String convertByteArrayToHexString(byte[] b) {
271       if (b != null) {
272         StringBuilder s = new StringBuilder(2 * b.length);
273
274         for (int i = 0; i < b.length; ++i) {
275           final String t = Integer.toHexString(b[i]);
276           final int l = t.length();
277           if (l > 2) {
278             s.append(t.substring(l - 2));
279           } else {
280             if (l == 1) {
281               s.append("0");
282             }
283             s.append(t);
284           }
```

Appendix B: Modifications to Google's Secure Element API Library

```
285        }
286
287        return s.toString();
288      } else {
289        return "";
290      }
291    }
```

```
302    public void authenticate(byte[] token) {
303      try {
304        Log.d(TAG,
305              "authenticate() for " +
306              ((mPackageName != null) ? mPackageName : "[null]") +
307              ": " + convertByteArrayToHexString(token));
308        sService.authenticate(mPackageName, token);
309      } catch (Exception e) {
310        attemptDeadServiceRecovery(e);
311      }
312    }
313  }
```

B.2 Class NfcExecutionEnvironment

```
1   /*
2    * Copyright (C) 2011 The Android Open Source Project
3    * Modifications (debug output): (C) 2012 Michael Roland
4    *
5    * Licensed under the Apache License, Version 2.0 (the
6    * "License"); you may not use this file except in compliance
7    * with the License. You may obtain a copy of the License at
8    *
9    *      http://www.apache.org/licenses/LICENSE-2.0
10   *
11   * Unless required by applicable law or agreed to in writing,
12   * software distributed under the License is distributed on
13   * an "AS IS" BASIS, WITHOUT WARRANTIES OR CONDITIONS OF ANY
14   * KIND, either express or implied. See the License for the
15   * specific language governing permissions and limitations
16   * under the License.
17   */
18
19  package com.android.nfc_extras;
20
21  import android.os.Binder;
22  import android.os.Bundle;
23  import android.os.RemoteException;
24  import android.util.Log;
25
26  import java.io.IOException;
27
28  public class NfcExecutionEnvironment {
29    private static final String TAG="NfcExecutionEnvironment";
30
31    private final NfcAdapterExtras mExtras;
32    private final Binder mToken;
```

```
46    public static final String ACTION_AID_SELECTED =
47            "com.android.nfc_extras.action.AID_SELECTED";


55    public static final String EXTRA_AID =
56            "com.android.nfc_extras.extra.AID";


70    public static final String ACTION_APDU_RECEIVED =
71            "com.android.nfc_extras.action.APDU_RECEIVED";


80    public static final String EXTRA_APDU_BYTES =
81            "com.android.nfc_extras.extra.APDU_BYTES";


88    public static final String ACTION_EMV_CARD_REMOVAL =
89            "com.android.nfc_extras.action.EMV_CARD_REMOVAL";


102   public static final String ACTION_MIFARE_ACCESS_DETECTED =
103           "com.android.nfc_extras.action.MIFARE_ACCESS_DETECTED";


113   public static final String EXTRA_MIFARE_BLOCK =
114           "com.android.nfc_extras.extra.MIFARE_BLOCK";
115
116   NfcExecutionEnvironment(NfcAdapterExtras extras) {
117     mExtras = extras;
118     mToken = new Binder();
119   }


135   public void open() throws IOException {
136     try {
137       Log.d(TAG,
138             "open() for " +
139             ((mExtras.mPackageName != null) ?
140                 mExtras.mPackageName : "[null]"));
141       Bundle b = mExtras.getService().open(mExtras.mPackageName,
142                                            mToken);
143       throwBundle(b);
144     } catch (Exception e) {
145       mExtras.attemptDeadServiceRecovery(e);
146       throw new IOException("NFC Service was dead, try again");
147     }
148   }


158   public void close() throws IOException {
159     try {
160       Log.d(TAG,
161             "close() for " +
162             ((mExtras.mPackageName != null) ?
163                 mExtras.mPackageName : "[null]"));
164       throwBundle(mExtras.getService().close(
165           mExtras.mPackageName,
166           mToken));
167     } catch (Exception e) {
```

Appendix B: Modifications to Google's Secure Element API Library

```java
168              mExtras.attemptDeadServiceRecovery(e);
169              throw new IOException("NFC Service was dead");
170          }
171      }
172
173      /**
174       * Convert a byte array into its hexadecimal string form.
175       * @param b   Byte array.
176       * @return    Hexadecimal string representation.
177       */
178      private static String convertByteArrayToHexString(byte[] b) {
179        if (b != null) {
180          StringBuilder s = new StringBuilder(2 * b.length);
181
182          for (int i = 0; i < b.length; ++i) {
183            final String t = Integer.toHexString(b[i]);
184            final int l = t.length();
185            if (l > 2) {
186              s.append(t.substring(l - 2));
187            } else {
188              if (l == 1) {
189                s.append("0");
190              }
191              s.append(t);
192            }
193          }
194
195          return s.toString();
196        } else {
197          return "";
198        }
199      }
```

```java
209      public byte[] transceive(byte[] in) throws IOException {
210        Bundle b;
211        try {
212          Log.d(TAG,
213                  "transceive() for " +
214                  ((mExtras.mPackageName != null) ?
215                       mExtras.mPackageName : "[null]") +
216                  ": C-APDU=" + convertByteArrayToHexString(in));
217          b = mExtras.getService().transceive(mExtras.mPackageName,
218                                               in);
219        } catch (Exception e) {
220          mExtras.attemptDeadServiceRecovery(e);
221          throw new IOException(
222              "NFC Service was dead, need to re-open");
223        }
224        throwBundle(b);
225        byte[] out = b.getByteArray("out");
226        Log.d(TAG,
227              "transceive() for " +
228              ((mExtras.mPackageName != null) ?
229                   mExtras.mPackageName : "[null]") +
230              ": R-APDU=" + convertByteArrayToHexString(out));
231        return out;
232      }
233
234      private static void throwBundle(Bundle b) throws IOException {
235        if (b.getInt("e") == -1) {
236          Log.d(TAG, "IOException: " + b.getString("m", "[null]"));
237          throw new IOException(b.getString("m"));
238        }
239      }
240    }
```

Index

A
Access control, 62, 104, 105, 108, 111, 112, 164
  bypass, 112
  enforcer, 110
Access Rule Applet, 111
Android, 2, 57, 107, 148
  vulnerabilities, 58
Answer-to-Reset, 14, 105, 110
Answer-to-Select, 17
Anti-collision, 16
APDU, *see* Application Protocol Data Unit
App, 1, 57, 120, 165
  signature, *see* Code signing
Applet, 18, 105, 142
Application processor, 55, 104, 112, 120, 141, 165
Application Protocol Data Unit, 14, 105, 106, 108, 111, 120, 131
ARA, *see* Access Rule Applet
ARM TrustZone, 59
ATR, *see* Answer-to-Reset
ATS, *see* Answer-to-Select
Authenticity, 79
Authorization, 79, 83
Automotive computer system, 37

C
CA, *see* Certification authority
Car
  immobilizer, 34
  key, 34–36
Card emulation, 2, 5, 21, 27, 40, 44, 62, 103, 164, 168
Card emulator, 121, 124, 128, 155, 165
Card Manager, 18, 115, 131, 149

Certificate, 81, 105, 109, 111
  content mapping, 81
  format, 89
  lifespan, 86, 164
Certification authority, 79
Chain model, 87
Chip & PIN, 29, 60
Cloning, 56
Cloud, 38
Code signing, 57, 85, 106, 107, 112
Collision resistance, 72
Command-response delay, 132, 137, 156
Common Criteria, 49
Compute Cryptographic Checksum, 153
Connection handover, 27, 33, 36, 38, 40, 42, 43, 82
Content spoofing, 52
Credential storage, 103
Crypto-1 cipher, 48

D
Data insertion, 47
Data manipulation, 71
Data modification, 47, 56
Denial-of-service, 62, 97, 115, 117, 142, 165
Digital signature, 53, 54, 62, 72, 166
  partial, 74, 82, 83, 97
  scope, 74, 78
  trust, 79, 91, 97, 100
Distance-bounding, 51

E
Eavesdropping, 43, 47
EMV, 28, 50, 54, 60, 123
  cloud-based, 142

EMV mode, *see* Chip & PIN
mag-stripe mode, 29, 151

F
FDT, *see* Frame delay time
Feature phone, 1
FeliCa, *see* JIS X 6319-4
Fleet management, 36
Frame delay time, 122
Frame waiting time, 123
Fuzzing, 53
FWT, *see* Frame waiting time

G
Get Processing Options, 152, 153
GlobalPlatform, 18, 42, 115, 116
Google Wallet, 5, 60, 108, 147, 148, 165
    on-card component, *see* On-card component
    PIN, 154, 157
    relay attack, 165
    unlock command, 154, 157

H
Hash function, 72
HCE, *see* Host-based card emulation
Host-based card emulation, 2, 21, 28, 56, 120, 125, 155, 168

I
In-browser payment, 54, 143
Inductive coupling, 15
ISO/IEC 14443, 15–17, 28, 122
ISO/IEC 15693, 17
ISO/IEC 18092, 20
ISO/IEC 7816, 13, 14, 18

J
Jail breaking, *see* Privilege escalation
Java Card, 18
JIS X 6319-4, 17

L
LLCP, *see* Logical Link Control Protocol
Logical Link Control Protocol, 20, 48

M
Mafia fraud, *see* Relay attack

MasterCard PayPass, 29, 60, 147, 151
MIFARE Classic, 48

N
NDEF, *see* NFC Data Exchange Format
Near Field Communication, 1, 19
    Card emulation mode, *see* Card emulation
    key, 34–36
    operating modes, 20
    Peer-to-peer mode, 20, 40, 42
    Reader/writer mode, 21, 40, 43, 106
    Record Type Definition, 24
    security, 3, 4, 47, 51, 163, 167
    tag, *see* Tagging
    wired interface, 28
NFC, *see* Near Field Communication
NFC Data Exchange Format, 22, 69
    API, 77
    chunk flag, 22, 24, 76
    connection handover, *see* Connection handover
    external type, 24
    message, 24
    parser, 77
    record, 22
    short record, 23, 75
    signature, 73, *see* aso Signature Record Type Definition
    smart poster record, *see* Smart poster
    text record, 25, 81
    type name format, 23, 76, 77, 93
    URI record, 25, 81
    well-known type, 24
NFC Forum, 3, 20, 54, 88, 166
Normalized form, 77, 78, 99

O
On-board credentials, 59
On-card component, 149, 150, 154
On-device secure payment, 158
Online signature generation, 85, 164
Open Mobile API, 109, 110
Over-the-air management, 42, 55, 104, 115, 141, 142

P
Pairing
    out-of-band, *see* Connection handover
PC/SC, *see* Personal Computer/Smart Card interface

Index

PCD, *see* Proximity Coupling Device
Personal Computer/Smart Card interface, 19, 131
Personalization, 34
PICC, *see* Proximity Integrated Circuit Card
PKI, *see* Public-key infrastructure
Power analysis, 49
Preimage resistence, 72
Private key, 38, 80, 81, 84, 85
Privilege escalation, 53, 57, 60, 114, 120, 147, 165
   framework, 58, 154
Proximity Coupling Device, 16
Proximity Integrated Circuit Card, 16
Public-key infrastructure, 79, 91, 99, 163, 167

R

Record composition attack, 62, 96, 97, 164
Record hiding, 93
Record joining, 93, 95
Relay attack, 5, 50, 56, 104, 115, 118
   countermeasures, 51, 141

S

Secret key, *see* Private key
Secure element, 4, 21, 27, 37, 40, 41, 44, 54, 103, 164, 168
   API, 62, 104, 127, 148
   attacks, 114
   cloud-based, 124, 168
   external mode, 40, 41, 130, 135
   internal mode, 40, 41, 104, 115, 118, 120, 130, 135, 141, 142, 158
Secure Element Evaluation Kit, 109
Security domain, 18, 105, 115
SEEK, *see* Secure Element Evaluation Kit
Shell model, 86
Signature Record Type Definition, 4, 54, 62, 73, 88, 163
   coverage, 90, 92
   weaknesses, 90, 98
SIM card, *see* Universal Integrated Circuit Card
Single Wire Protocol, 28
Skimming, 55, 104
Smart phone, 1
   app, *see* App
   security, 57
Smart poster, 26, 52, 69, 71, 81, 83, 97
Smartcard, 4, 13, 27, 40, 103, 106
   attack, 49
   file system, 14
   lifecycle, 116
   protocol stack, 14
   security, 44
Soft-SE, *see* Host-based card emulation
Software card emulation, *see* Host-based card emulation
Software-based relay attack, 62, 115, 118, 120, 147, 154, 165
   prototype, 126
   solutions, 157
Spoofing, 53, 70, 72
Subscriber Identity Module, *see* Universal Integrated Circuit Card
SWP, *see* Single Wire Protocol

T

Tag-length-value format, 19
Tagging, 4, 5, 21, 33, 40, 43, 52, 61, 69, 163
   security, 52, 70
TLV, *see* Tag-length-value format
Trusted computing, 59, 141, 168
Trusted service manager, 42, 104
TrustZone, *see* ARM TrustZone
Two-factor authentication, 142

U

UICC, *see* Universal Integrated Circuit Card
Universal Integrated Circuit Card, 13, 27, 42, 105, 115
Usage data, 92, 164

W

Wallet
   digital, 41, 59
   Google, *see* Google Wallet
Wormhole attack, 50
Write protection, 52, 71

Z

Zapper, 72

Printed by Books on Demand, Germany